Nuclear Weapons: A Very Short Introduction

VERY SHORT INTRODUCTIONS are for anyone wanting a stimulating and accessible way into a new subject. They are written by experts, and have been translated into more than 45 different languages.

The series began in 1995, and now covers a wide variety of topics in every discipline. The VSI library currently contains over 650 volumes—a Very Short Introduction to everything from Psychology and Philosophy of Science to American History and Relativity—and continues to grow in every subject area.

Very Short Introductions available now:

ANIMAL BEHAVIOUR
Tristram D. Wyatt
THE ANIMAL KINGDOM
Peter Holland
ANIMAL RIGHTS David DeGrazia
THE ANTARCTIC Klaus Dodds
ANTHROPOCENE Erle C. Ellis
ANTISEMITISM Steven Beller
ANXIETY Daniel Freeman and
Jason Freeman
THE APOCRYPHAL GOSPELS
Paul Foster
APPLIED MATHEMATICS
Alain Goriely
ARCHAEOLOGY Paul Bahn
ARCHITECTURE Andrew Ballantyne
ARISTOCRACY William Doyle
ARISTOTLE Jonathan Barnes
ART HISTORY Dana Arnold
ART THEORY Cynthia Freeland
ARTIFICIAL INTELLIGENCE
Margaret A. Boden
ASIAN AMERICAN HISTORY
Madeline Y. Hsu
ASTROBIOLOGY David C. Catling
ASTROPHYSICS James Binney
ATHEISM Julian Baggini
THE ATMOSPHERE Paul I. Palmer
AUGUSTINE Henry Chadwick
AUSTRALIA Kenneth Morgan
AUTISM Uta Frith
AUTOBIOGRAPHY Laura Marcus
THE AVANT GARDE David Cottington
THE AZTECS Davíd Carrasco
BABYLONIA Trevor Bryce
BACTERIA Sebastian G. B. Amyes
BANKING John Goddard and
John O. S. Wilson
BARTHES Jonathan Culler
THE BEATS David Sterritt
BEAUTY Roger Scruton
BEHAVIOURAL ECONOMICS
Michelle Baddeley
BESTSELLERS John Sutherland
THE BIBLE John Riches
BIBLICAL ARCHAEOLOGY
Eric H. Cline
BIG DATA Dawn E. Holmes
BIOGEOGRAPHY Mark V. Lomolino
BIOGRAPHY Hermione Lee
BIOMETRICS Michael Fairhurst
BLACK HOLES Katherine Blundell
BLOOD Chris Cooper
THE BLUES Elijah Wald
THE BODY Chris Shilling
THE BOOK OF COMMON PRAYER
Brian Cummings
THE BOOK OF MORMON
Terryl Givens
BORDERS Alexander C. Diener and
Joshua Hagen
THE BRAIN Michael O'Shea
BRANDING Robert Jones
THE BRICS Andrew F. Cooper
THE BRITISH CONSTITUTION
Martin Loughlin
THE BRITISH EMPIRE Ashley Jackson
BRITISH POLITICS Tony Wright
BUDDHA Michael Carrithers
BUDDHISM Damien Keown
BUDDHIST ETHICS Damien Keown
BYZANTIUM Peter Sarris
C. S. LEWIS James Como
CALVINISM Jon Balserak
CANADA Donald Wright
CANCER Nicholas James
CAPITALISM James Fulcher
CATHOLICISM Gerald O'Collins
CAUSATION Stephen Mumford and
Rani Lill Anjum
THE CELL Terence Allen and
Graham Cowling
THE CELTS Barry Cunliffe
CHAOS Leonard Smith
CHARLES DICKENS Jenny Hartley
CHEMISTRY Peter Atkins
CHILD PSYCHOLOGY Usha Goswami
CHILDREN'S LITERATURE
Kimberley Reynolds
CHINESE LITERATURE Sabina Knight
CHOICE THEORY Michael Allingham
CHRISTIAN ART Beth Williamson
CHRISTIAN ETHICS D. Stephen Long
CHRISTIANITY Linda Woodhead
CIRCADIAN RHYTHMS Russell
Foster and Leon Kreitzman
CITIZENSHIP Richard Bellamy
CIVIL ENGINEERING
David Muir Wood
CLASSICAL LITERATURE William Allan
CLASSICAL MYTHOLOGY
Helen Morales

CLASSICS Mary Beard and John Henderson
CLAUSEWITZ Michael Howard
CLIMATE Mark Maslin
CLIMATE CHANGE Mark Maslin
CLINICAL PSYCHOLOGY Susan Llewelyn and Katie Aafjes-van Doorn
COGNITIVE NEUROSCIENCE Richard Passingham
THE COLD WAR Robert McMahon
COLONIAL AMERICA Alan Taylor
COLONIAL LATIN AMERICAN LITERATURE Rolena Adorno
COMBINATORICS Robin Wilson
COMEDY Matthew Bevis
COMMUNISM Leslie Holmes
COMPARATIVE LITERATURE Ben Hutchinson
COMPLEXITY John H. Holland
THE COMPUTER Darrel Ince
COMPUTER SCIENCE Subrata Dasgupta
CONCENTRATION CAMPS Dan Stone
CONFUCIANISM Daniel K. Gardner
THE CONQUISTADORS Matthew Restall and Felipe Fernández-Armesto
CONSCIENCE Paul Strohm
CONSCIOUSNESS Susan Blackmore
CONTEMPORARY ART Julian Stallabrass
CONTEMPORARY FICTION Robert Eaglestone
CONTINENTAL PHILOSOPHY Simon Critchley
COPERNICUS Owen Gingerich
CORAL REEFS Charles Sheppard
CORPORATE SOCIAL RESPONSIBILITY Jeremy Moon
CORRUPTION Leslie Holmes
COSMOLOGY Peter Coles
COUNTRY MUSIC Richard Carlin
CRIME FICTION Richard Bradford
CRIMINAL JUSTICE Julian V. Roberts
CRIMINOLOGY Tim Newburn
CRITICAL THEORY Stephen Eric Bronner
THE CRUSADES Christopher Tyerman
CRYPTOGRAPHY Fred Piper and Sean Murphy
CRYSTALLOGRAPHY A. M. Glazer
THE CULTURAL REVOLUTION Richard Curt Kraus
DADA AND SURREALISM David Hopkins
DANTE Peter Hainsworth and David Robey
DARWIN Jonathan Howard
THE DEAD SEA SCROLLS Timothy H. Lim
DECADENCE David Weir
DECOLONIZATION Dane Kennedy
DEMENTIA Kathleen Taylor
DEMOCRACY Bernard Crick
DEMOGRAPHY Sarah Harper
DEPRESSION Jan Scott and Mary Jane Tacchi
DERRIDA Simon Glendinning
DESCARTES Tom Sorell
DESERTS Nick Middleton
DESIGN John Heskett
DEVELOPMENT Ian Goldin
DEVELOPMENTAL BIOLOGY Lewis Wolpert
THE DEVIL Darren Oldridge
DIASPORA Kevin Kenny
DICTIONARIES Lynda Mugglestone
DINOSAURS David Norman
DIPLOMACY Joseph M. Siracusa
DOCUMENTARY FILM Patricia Aufderheide
DREAMING J. Allan Hobson
DRUGS Les Iversen
DRUIDS Barry Cunliffe
DYNASTY Jeroen Duindam
DYSLEXIA Margaret J. Snowling
EARLY MUSIC Thomas Forrest Kelly
THE EARTH Martin Redfern
EARTH SYSTEM SCIENCE Tim Lenton
ECONOMICS Partha Dasgupta
EDUCATION Gary Thomas
EGYPTIAN MYTH Geraldine Pinch
EIGHTEENTH-CENTURY BRITAIN Paul Langford
THE ELEMENTS Philip Ball
ÉMILE ZOLA Brian Nelson
EMOTION Dylan Evans
EMPIRE Stephen Howe
ENERGY SYSTEMS Nick Jenkins
ENGELS Terrell Carver
ENGINEERING David Blockley

THE ENGLISH LANGUAGE Simon Horobin
ENGLISH LITERATURE Jonathan Bate
THE ENLIGHTENMENT John Robertson
ENTREPRENEURSHIP Paul Westhead and Mike Wright
ENVIRONMENTAL ECONOMICS Stephen Smith
ENVIRONMENTAL ETHICS Robin Attfield
ENVIRONMENTAL LAW Elizabeth Fisher
ENVIRONMENTAL POLITICS Andrew Dobson
EPICUREANISM Catherine Wilson
EPIDEMIOLOGY Rodolfo Saracci
ETHICS Simon Blackburn
ETHNOMUSICOLOGY Timothy Rice
THE ETRUSCANS Christopher Smith
EUGENICS Philippa Levine
THE EUROPEAN UNION Simon Usherwood and John Pinder
EUROPEAN UNION LAW Anthony Arnull
EVOLUTION Brian and Deborah Charlesworth
EXISTENTIALISM Thomas Flynn
EXPLORATION Stewart A. Weaver
EXTINCTION Paul B. Wignall
THE EYE Michael Land
FAIRY TALE Marina Warner
FAMILY LAW Jonathan Herring
FASCISM Kevin Passmore
FASHION Rebecca Arnold
FEDERALISM Mark J. Rozell and Clyde Wilcox
FEMINISM Margaret Walters
FILM Michael Wood
FILM MUSIC Kathryn Kalinak
FILM NOIR James Naremore
FIRE Andrew C. Scott
THE FIRST WORLD WAR Michael Howard
FOLK MUSIC Mark Slobin
FOOD John Krebs
FORENSIC PSYCHOLOGY David Canter
FORENSIC SCIENCE Jim Fraser
FORESTS Jaboury Ghazoul
FOSSILS Keith Thomson
FOUCAULT Gary Gutting
THE FOUNDING FATHERS R. B. Bernstein
FRACTALS Kenneth Falconer
FREE SPEECH Nigel Warburton
FREE WILL Thomas Pink
FREEMASONRY Andreas Önnerfors
FRENCH LITERATURE John D. Lyons
FRENCH PHILOSOPHY Stephen Gaukroger and Knox Peden
THE FRENCH REVOLUTION William Doyle
FREUD Anthony Storr
FUNDAMENTALISM Malise Ruthven
FUNGI Nicholas P. Money
THE FUTURE Jennifer M. Gidley
GALAXIES John Gribbin
GALILEO Stillman Drake
GAME THEORY Ken Binmore
GANDHI Bhikhu Parekh
GARDEN HISTORY Gordon Campbell
GENES Jonathan Slack
GENIUS Andrew Robinson
GENOMICS John Archibald
GEOFFREY CHAUCER David Wallace
GEOGRAPHY John Matthews and David Herbert
GEOLOGY Jan Zalasiewicz
GEOPHYSICS William Lowrie
GEOPOLITICS Klaus Dodds
GEORGE BERNARD SHAW Christopher Wixson
GERMAN LITERATURE Nicholas Boyle
GERMAN PHILOSOPHY Andrew Bowie
THE GHETTO Bryan Cheyette
GLACIATION David J. A. Evans
GLOBAL CATASTROPHES Bill McGuire
GLOBAL ECONOMIC HISTORY Robert C. Allen
GLOBALIZATION Manfred Steger
GOD John Bowker
GOETHE Ritchie Robertson
THE GOTHIC Nick Groom
GOVERNANCE Mark Bevir
GRAVITY Timothy Clifton
THE GREAT DEPRESSION AND THE NEW DEAL Eric Rauchway
HABERMAS James Gordon Finlayson
THE HABSBURG EMPIRE Martyn Rady

HAPPINESS Daniel M. Haybron
THE HARLEM RENAISSANCE Cheryl A. Wall
THE HEBREW BIBLE AS LITERATURE Tod Linafelt
HEGEL Peter Singer
HEIDEGGER Michael Inwood
THE HELLENISTIC AGE Peter Thonemann
HEREDITY John Waller
HERMENEUTICS Jens Zimmermann
HERODOTUS Jennifer T. Roberts
HIEROGLYPHS Penelope Wilson
HINDUISM Kim Knott
HISTORY John H. Arnold
THE HISTORY OF ASTRONOMY Michael Hoskin
THE HISTORY OF CHEMISTRY William H. Brock
THE HISTORY OF CHILDHOOD James Marten
THE HISTORY OF CINEMA Geoffrey Nowell-Smith
THE HISTORY OF LIFE Michael Benton
THE HISTORY OF MATHEMATICS Jacqueline Stedall
THE HISTORY OF MEDICINE William Bynum
THE HISTORY OF PHYSICS J. L. Heilbron
THE HISTORY OF TIME Leofranc Holford-Strevens
HIV AND AIDS Alan Whiteside
HOBBES Richard Tuck
HOLLYWOOD Peter Decherney
THE HOLY ROMAN EMPIRE Joachim Whaley
HOME Michael Allen Fox
HOMER Barbara Graziosi
HORMONES Martin Luck
HUMAN ANATOMY Leslie Klenerman
HUMAN EVOLUTION Bernard Wood
HUMAN RIGHTS Andrew Clapham
HUMANISM Stephen Law
HUME A. J. Ayer
HUMOUR Noël Carroll
THE ICE AGE Jamie Woodward
IDENTITY Florian Coulmas
IDEOLOGY Michael Freeden
THE IMMUNE SYSTEM Paul Klenerman
INDIAN CINEMA Ashish Rajadhyaksha
INDIAN PHILOSOPHY Sue Hamilton
THE INDUSTRIAL REVOLUTION Robert C. Allen
INFECTIOUS DISEASE Marta L. Wayne and Benjamin M. Bolker
INFINITY Ian Stewart
INFORMATION Luciano Floridi
INNOVATION Mark Dodgson and David Gann
INTELLECTUAL PROPERTY Siva Vaidhyanathan
INTELLIGENCE Ian J. Deary
INTERNATIONAL LAW Vaughan Lowe
INTERNATIONAL MIGRATION Khalid Koser
INTERNATIONAL RELATIONS Christian Reus-Smit
INTERNATIONAL SECURITY Christopher S. Browning
IRAN Ali M. Ansari
ISLAM Malise Ruthven
ISLAMIC HISTORY Adam Silverstein
ISOTOPES Rob Ellam
ITALIAN LITERATURE Peter Hainsworth and David Robey
JESUS Richard Bauckham
JEWISH HISTORY David N. Myers
JOURNALISM Ian Hargreaves
JUDAISM Norman Solomon
JUNG Anthony Stevens
KABBALAH Joseph Dan
KAFKA Ritchie Robertson
KANT Roger Scruton
KEYNES Robert Skidelsky
KIERKEGAARD Patrick Gardiner
KNOWLEDGE Jennifer Nagel
THE KORAN Michael Cook
KOREA Michael J. Seth
LAKES Warwick F. Vincent
LANDSCAPE ARCHITECTURE Ian H. Thompson
LANDSCAPES AND GEOMORPHOLOGY Andrew Goudie and Heather Viles
LANGUAGES Stephen R. Anderson
LATE ANTIQUITY Gillian Clark

LAW Raymond Wacks
THE LAWS OF THERMODYNAMICS
Peter Atkins
LEADERSHIP Keith Grint
LEARNING Mark Haselgrove
LEIBNIZ Maria Rosa Antognazza
LEO TOLSTOY Liza Knapp
LIBERALISM Michael Freeden
LIGHT Ian Walmsley
LINCOLN Allen C. Guelzo
LINGUISTICS Peter Matthews
LITERARY THEORY Jonathan Culler
LOCKE John Dunn
LOGIC Graham Priest
LOVE Ronald de Sousa
MACHIAVELLI Quentin Skinner
MADNESS Andrew Scull
MAGIC Owen Davies
MAGNA CARTA Nicholas Vincent
MAGNETISM Stephen Blundell
MALTHUS Donald Winch
MAMMALS T. S. Kemp
MANAGEMENT John Hendry
MAO Delia Davin
MARINE BIOLOGY Philip V. Mladenov
THE MARQUIS DE SADE John Phillips
MARTIN LUTHER Scott H. Hendrix
MARTYRDOM Jolyon Mitchell
MARX Peter Singer
MATERIALS Christopher Hall
MATHEMATICAL FINANCE
Mark H. A. Davis
MATHEMATICS Timothy Gowers
MATTER Geoff Cottrell
THE MEANING OF LIFE
Terry Eagleton
MEASUREMENT David Hand
MEDICAL ETHICS Michael Dunn and
Tony Hope
MEDICAL LAW Charles Foster
MEDIEVAL BRITAIN John Gillingham
and Ralph A. Griffiths
MEDIEVAL LITERATURE
Elaine Treharne
MEDIEVAL PHILOSOPHY
John Marenbon
MEMORY Jonathan K. Foster
METAPHYSICS Stephen Mumford
METHODISM William J. Abraham
THE MEXICAN REVOLUTION
Alan Knight
MICHAEL FARADAY
Frank A. J. L. James
MICROBIOLOGY Nicholas P. Money
MICROECONOMICS Avinash Dixit
MICROSCOPY Terence Allen
THE MIDDLE AGES Miri Rubin
MILITARY JUSTICE Eugene R. Fidell
MILITARY STRATEGY
Antulio J. Echevarria II
MINERALS David Vaughan
MIRACLES Yujin Nagasawa
MODERN ARCHITECTURE
Adam Sharr
MODERN ART David Cottington
MODERN CHINA Rana Mitter
MODERN DRAMA
Kirsten E. Shepherd-Barr
MODERN FRANCE
Vanessa R. Schwartz
MODERN INDIA Craig Jeffrey
MODERN IRELAND Senia Pašeta
MODERN ITALY Anna Cento Bull
MODERN JAPAN
Christopher Goto-Jones
MODERN LATIN AMERICAN
LITERATURE
Roberto González Echevarría
MODERN WAR Richard English
MODERNISM Christopher Butler
MOLECULAR BIOLOGY Aysha Divan
and Janice A. Royds
MOLECULES Philip Ball
MONASTICISM Stephen J. Davis
THE MONGOLS Morris Rossabi
MOONS David A. Rothery
MORMONISM
Richard Lyman Bushman
MOUNTAINS Martin F. Price
MUHAMMAD Jonathan A. C. Brown
MULTICULTURALISM Ali Rattansi
MULTILINGUALISM John C. Maher
MUSIC Nicholas Cook
MYTH Robert A. Segal
NAPOLEON David Bell
THE NAPOLEONIC WARS
Mike Rapport
NATIONALISM Steven Grosby
NATIVE AMERICAN LITERATURE
Sean Teuton
NAVIGATION Jim Bennett
NAZI GERMANY Jane Caplan

NELSON MANDELA Elleke Boehmer
NEOLIBERALISM Manfred Steger and Ravi Roy
NETWORKS Guido Caldarelli and Michele Catanzaro
THE NEW TESTAMENT Luke Timothy Johnson
THE NEW TESTAMENT AS LITERATURE Kyle Keefer
NEWTON Robert Iliffe
NIELS BOHR J. L. Heilbron
NIETZSCHE Michael Tanner
NINETEENTH-CENTURY BRITAIN Christopher Harvie and H. C. G. Matthew
THE NORMAN CONQUEST George Garnett
NORTH AMERICAN INDIANS Theda Perdue and Michael D. Green
NORTHERN IRELAND Marc Mulholland
NOTHING Frank Close
NUCLEAR PHYSICS Frank Close
NUCLEAR POWER Maxwell Irvine
NUCLEAR WEAPONS Joseph M. Siracusa
NUMBER THEORY Robin Wilson
NUMBERS Peter M. Higgins
NUTRITION David A. Bender
OBJECTIVITY Stephen Gaukroger
OCEANS Dorrik Stow
THE OLD TESTAMENT Michael D. Coogan
THE ORCHESTRA D. Kern Holoman
ORGANIC CHEMISTRY Graham Patrick
ORGANIZATIONS Mary Jo Hatch
ORGANIZED CRIME Georgios A. Antonopoulos and Georgios Papanicolaou
ORTHODOX CHRISTIANITY A. Edward Siecienski
PAGANISM Owen Davies
PAIN Rob Boddice
THE PALESTINIAN-ISRAELI CONFLICT Martin Bunton
PANDEMICS Christian W. McMillen
PARTICLE PHYSICS Frank Close
PAUL E. P. Sanders
PEACE Oliver P. Richmond
PENTECOSTALISM William K. Kay
PERCEPTION Brian Rogers
THE PERIODIC TABLE Eric R. Scerri
PHILOSOPHY Edward Craig
PHILOSOPHY IN THE ISLAMIC WORLD Peter Adamson
PHILOSOPHY OF BIOLOGY Samir Okasha
PHILOSOPHY OF LAW Raymond Wacks
PHILOSOPHY OF SCIENCE Samir Okasha
PHILOSOPHY OF RELIGION Tim Bayne
PHOTOGRAPHY Steve Edwards
PHYSICAL CHEMISTRY Peter Atkins
PHYSICS Sidney Perkowitz
PILGRIMAGE Ian Reader
PLAGUE Paul Slack
PLANETS David A. Rothery
PLANTS Timothy Walker
PLATE TECTONICS Peter Molnar
PLATO Julia Annas
POETRY Bernard O'Donoghue
POLITICAL PHILOSOPHY David Miller
POLITICS Kenneth Minogue
POPULISM Cas Mudde and Cristóbal Rovira Kaltwasser
POSTCOLONIALISM Robert Young
POSTMODERNISM Christopher Butler
POSTSTRUCTURALISM Catherine Belsey
POVERTY Philip N. Jefferson
PREHISTORY Chris Gosden
PRESOCRATIC PHILOSOPHY Catherine Osborne
PRIVACY Raymond Wacks
PROBABILITY John Haigh
PROGRESSIVISM Walter Nugent
PROHIBITION W. J. Rorabaugh
PROJECTS Andrew Davies
PROTESTANTISM Mark A. Noll
PSYCHIATRY Tom Burns
PSYCHOANALYSIS Daniel Pick
PSYCHOLOGY Gillian Butler and Freda McManus
PSYCHOLOGY OF MUSIC Elizabeth Hellmuth Margulis
PSYCHOPATHY Essi Viding
PSYCHOTHERAPY Tom Burns and Eva Burns-Lundgren

PUBLIC ADMINISTRATION Stella Z. Theodoulou and Ravi K. Roy
PUBLIC HEALTH Virginia Berridge
PURITANISM Francis J. Bremer
THE QUAKERS Pink Dandelion
QUANTUM THEORY John Polkinghorne
RACISM Ali Rattansi
RADIOACTIVITY Claudio Tuniz
RASTAFARI Ennis B. Edmonds
READING Belinda Jack
THE REAGAN REVOLUTION Gil Troy
REALITY Jan Westerhoff
RECONSTRUCTION Allen C. Guelzo
THE REFORMATION Peter Marshall
RELATIVITY Russell Stannard
RELIGION IN AMERICA Timothy Beal
THE RENAISSANCE Jerry Brotton
RENAISSANCE ART Geraldine A. Johnson
RENEWABLE ENERGY Nick Jelley
REPTILES T. S. Kemp
REVOLUTIONS Jack A. Goldstone
RHETORIC Richard Toye
RISK Baruch Fischhoff and John Kadvany
RITUAL Barry Stephenson
RIVERS Nick Middleton
ROBOTICS Alan Winfield
ROCKS Jan Zalasiewicz
ROMAN BRITAIN Peter Salway
THE ROMAN EMPIRE Christopher Kelly
THE ROMAN REPUBLIC David M. Gwynn
ROMANTICISM Michael Ferber
ROUSSEAU Robert Wokler
RUSSELL A. C. Grayling
THE RUSSIAN ECONOMY Richard Connolly
RUSSIAN HISTORY Geoffrey Hosking
RUSSIAN LITERATURE Catriona Kelly
THE RUSSIAN REVOLUTION S. A. Smith
THE SAINTS Simon Yarrow
SAVANNAS Peter A. Furley
SCEPTICISM Duncan Pritchard
SCHIZOPHRENIA Chris Frith and Eve Johnstone
SCHOPENHAUER Christopher Janaway
SCIENCE AND RELIGION Thomas Dixon
SCIENCE FICTION David Seed
THE SCIENTIFIC REVOLUTION Lawrence M. Principe
SCOTLAND Rab Houston
SECULARISM Andrew Copson
SEXUAL SELECTION Marlene Zuk and Leigh W. Simmons
SEXUALITY Véronique Mottier
SHAKESPEARE'S COMEDIES Bart van Es
SHAKESPEARE'S SONNETS AND POEMS Jonathan F. S. Post
SHAKESPEARE'S TRAGEDIES Stanley Wells
SIKHISM Eleanor Nesbitt
THE SILK ROAD James A. Millward
SLANG Jonathon Green
SLEEP Steven W. Lockley and Russell G. Foster
SMELL Matthew Cobb
SOCIAL AND CULTURAL ANTHROPOLOGY John Monaghan and Peter Just
SOCIAL PSYCHOLOGY Richard J. Crisp
SOCIAL WORK Sally Holland and Jonathan Scourfield
SOCIALISM Michael Newman
SOCIOLINGUISTICS John Edwards
SOCIOLOGY Steve Bruce
SOCRATES C. C. W. Taylor
SOUND Mike Goldsmith
SOUTHEAST ASIA James R. Rush
THE SOVIET UNION Stephen Lovell
THE SPANISH CIVIL WAR Helen Graham
SPANISH LITERATURE Jo Labanyi
SPINOZA Roger Scruton
SPIRITUALITY Philip Sheldrake
SPORT Mike Cronin
STARS Andrew King
STATISTICS David J. Hand
STEM CELLS Jonathan Slack
STOICISM Brad Inwood
STRUCTURAL ENGINEERING David Blockley
STUART BRITAIN John Morrill
THE SUN Philip Judge
SUPERCONDUCTIVITY Stephen Blundell
SUPERSTITION Stuart Vyse
SYMMETRY Ian Stewart

SYNAESTHESIA Julia Simner
SYNTHETIC BIOLOGY Jamie A. Davies
SYSTEMS BIOLOGY Eberhard O. Voit
TAXATION Stephen Smith
TEETH Peter S. Ungar
TELESCOPES Geoff Cottrell
TERRORISM Charles Townshend
THEATRE Marvin Carlson
THEOLOGY David F. Ford
THINKING AND REASONING Jonathan St B. T. Evans
THOMAS AQUINAS Fergus Kerr
THOUGHT Tim Bayne
TIBETAN BUDDHISM Matthew T. Kapstein
TIDES David George Bowers and Emyr Martyn Roberts
TOCQUEVILLE Harvey C. Mansfield
TOPOLOGY Richard Earl
TRAGEDY Adrian Poole
TRANSLATION Matthew Reynolds
THE TREATY OF VERSAILLES Michael S. Neiberg
TRIGONOMETRY Glen Van Brummelen
THE TROJAN WAR Eric H. Cline
TRUST Katherine Hawley
THE TUDORS John Guy
TWENTIETH-CENTURY BRITAIN Kenneth O. Morgan
TYPOGRAPHY Paul Luna
THE UNITED NATIONS Jussi M. Hanhimäki
UNIVERSITIES AND COLLEGES David Palfreyman and Paul Temple
THE U.S. CIVIL WAR Louis P. Masur
THE U.S. CONGRESS Donald A. Ritchie
THE U.S. CONSTITUTION David J. Bodenhamer
THE U.S. SUPREME COURT Linda Greenhouse
UTILITARIANISM Katarzyna de Lazari-Radek and Peter Singer
UTOPIANISM Lyman Tower Sargent
VETERINARY SCIENCE James Yeates
THE VIKINGS Julian D. Richards
VIRUSES Dorothy H. Crawford
VOLTAIRE Nicholas Cronk
WAR AND TECHNOLOGY Alex Roland
WATER John Finney
WAVES Mike Goldsmith
WEATHER Storm Dunlop
THE WELFARE STATE David Garland
WILLIAM SHAKESPEARE Stanley Wells
WITCHCRAFT Malcolm Gaskill
WITTGENSTEIN A. C. Grayling
WORK Stephen Fineman
WORLD MUSIC Philip Bohlman
THE WORLD TRADE ORGANIZATION Amrita Narlikar
WORLD WAR II Gerhard L. Weinberg
WRITING AND SCRIPT Andrew Robinson
ZIONISM Michael Stanislawski

Available soon:

OVIDLLEWELYN Morgan
ECOLOGY Jaboury Ghazoul
MODERN BRAZIL Anthony W. Pereira
SOFT MATTERTOM McLeish
PHILOSOPHICAL METHOD Timothy Williamson

Joseph M. Siracusa

NUCLEAR WEAPONS

A Very Short Introduction

THIRD EDITION

OXFORD UNIVERSITY PRESS

OXFORD
UNIVERSITY PRESS

Great Clarendon Street, Oxford, OX2 6DP,
United Kingdom

Oxford University Press is a department of the University of Oxford.
It furthers the University's objective of excellence in research, scholarship,
and education by publishing worldwide. Oxford is a registered trade mark of
Oxford University Press in the UK and in certain other countries

First edition published 2005
Second edition published 2015
Third edition published 2020

Published in the United States of America by Oxford University Press
198 Madison Avenue, New York, NY 10016, United States of America

British Library Cataloguing in Publication Data
Data available

Library of Congress Control Number: 2020935142

ISBN 978-0-19-886053-2

Printed and bound by
CPI Group (UK) Ltd, Croydon, CR0 4YY

for my wife Candice
CXVI

Contents

Preface

While underscoring the theme that nothing endangers the world more than nuclear weapons, the principal aim of this revision has been to recount the historical development of nuclear weapons and the policies they have generated since the end of the Cold War (see Chapter 7). Still focusing on the most important, common, and recurring questions about nuclear weapons, the discussion continues to rest on a single premise: the bomb still matters.

Nuclear weapons have not been used in anger since Hiroshima and Nagasaki, nearly seventy-five years ago, yet real concerns about their potential use have remained conspicuously present on the global stage. 'If you introduce them,' observed Representative Adam Smith, chairman of the US House Services Committee, in 2018, 'you cannot predict what your adversaries are going to counter with, and an all-out nuclear war is the likely result, with the complete destruction of the planet.'

The end of Moscow-dominated communism in 1991 did little to solve the problem of living dangerously with nuclear weapons. The nuclear past was never dead, nor was it even past. As President Bill Clinton's first Secretary of Defense Les Aspin aptly put it: 'The Cold War is over, the Soviet Union is no more. But the post-Cold War world is decidedly not post-nuclear.' For all the efforts to reduce nuclear stockpiles to zero, the bomb is here to stay. Gone may be the days when living with the bomb meant, in the words of former Secretary of State Madeleine Albright, 'Each night we knew that within minutes, perhaps through a misunderstanding, our world could end and morning never come', but if the threat of global thermonuclear war has receded, it has not disappeared. By any standard, the prospect of a global post-nuclear age has not progressed much further than wishful thinking.

Nuclear threats remain fundamental to relations between many states and threaten to become more and more important. The spread of nuclear weapons will likely spawn two potentially calamitous effects. The first is the possibility that terrorists will get their hands on nuclear weapons, something that has come into stark relief since the events of 9/11. To be sure, terrorists have not yet succeeded in initiating a nuclear attack. But, according to nuclear analysts, it's not because they can't. With a small quantity of enriched uranium, a handful of military supplies readily available on the Internet, and a small team of dedicated terrorists, they could potentially assemble a nuclear weapon in a matter of months, and deliver it by air, sea, rail, or road. The impact of such an attack in the heart of New York or London is almost unimaginable.

A second effect of the spread of nuclear weapons will be the proliferation of threats to use them, which will greatly complicate global security and in many respects be harder to undo. As more states join the nuclear club to enhance their prestige or overcome perceived insecurity, they will undergo their own nuclear learning curve, a process for which, as the experience of the nuclear states over the past seventy-five years has shown, there is no guarantee of success. The likelihood of mishaps along the way is only too real.

When the atomic bomb was unleashed on the mainland of Japan, in August 1945, in the closing stages of World War II, it was immediately apparent that this was not just another weapon (although it was that, too, as the A-bomb proved more efficient than 220 B-29s carrying 1,200 tons of incendiary bombs, 400 tons of high explosive bombs, and 500 tons of antipersonnel fragmentation bombs). In many respects, then, Hiroshima was not the kind of watershed moment that can only be seen in retrospect. President Harry S. Truman described the event to a startled world as the very 'harnessing of the basic power of the universe'. It was a view widely shared by influential atomic scientists.

Seven years later, in 1952, the United States scaled the nuclear ladder, detonating its first thermonuclear device in the Pacific. 'Mike', as the bomb was designated, exploded with a force 500 times greater than the bomb detonated over Hiroshima, in the process wiping the test island off the map. The H-bomb really changed everything, transforming the very nature of war and peace. Or, as Winston Churchill put it, 'The atomic bomb, with all its terror, did not carry us outside the scope of human control or manageable events, in thought or action, in peace or war. But... [with] the hydrogen bomb, the entire foundation of human affairs was revolutionized.' Indeed, it was a brave new world.

A sample of statistics from the nuclear age that followed provides a sobering reminder of the scale of the problem. Upwards of 128,000 nuclear weapons have been produced in the past seventy-five years, of which about 98 per cent were produced by the United States and the former Soviet Union. The nine current members of the nuclear club—the United States, Russia, the United Kingdom, France, India, Pakistan, China, Israel, and North Korea—still possess about 13,890 operational nuclear warheads between them. At least another fifteen countries currently have on hand enough highly enriched uranium for a nuclear weapon, while the same number of countries already have the delivery or ballistic missile systems to deliver them.

Within this context, the book looks at the science of nuclear weapons and how they differ from conventional weapons; the race to beat Nazi scientists to the bomb; the history of early attempts to control the bomb, through to the Soviet detonation of an atomic device in August 1949; the race to acquire the H-bomb, with its revolutionary implications; the history and politics of nuclear deterrence and arms control, against the backdrop of a changing global landscape, from the Cold War to the present; the prospect and promise of missile defence, from the end of World War II through to Ronald Reagan's dream of shielding the American homeland from a massive Soviet ballistic attack ('Star Wars'), through to Washington's post-Cold War reduced goal of defending against a small number of ballistic missiles (National Missile Defense) launched by a rogue state; and, finally, the historical development of nuclear weapons and the policies they have generated since the end of the Cold War.

Professor Joseph M. Siracusa
The Royal Melbourne Institute of Technology University
Melbourne, Australia

List of illustrations

The publisher and the author apologize for any errors or omissions in the above list. If contacted they will be pleased to rectify these at the earliest opportunity.

Chapter 1
What are nuclear weapons?

In 1951 the newly established US Federal Civil Defense Administration (FCDA) commissioned the production of a film to instruct children how to react in the event of a nuclear attack. The result was *Duck and Cover*, a film lasting nine minutes that was shown in schools throughout the United States during the 1950s and beyond. It featured a cartoon character, Bert the Turtle, who 'was very alert' and 'knew just what to do: duck and cover'. At the sound of an alarm or the flash of a brilliant light signalling a nuclear explosion, Bert would instantly tuck his body under his shell. It looked simple enough. And everyone loved the turtle.

Other FCDA initiatives of the early 1950s led to the creation of the Emergency Broadcast System, food stockpiles, civil defence classes, and public and private bomb shelters. The FCDA commissioned other civil defence films, but *Duck and Cover* became the most famous of the genre (Figure 1). In 2004 the US Library of Congress even included it in the National Film Registry of 'culturally, historically or aesthetically' significant motion pictures, a distinction it now shares with such feature film classics as *Birth of a Nation*, *Casablanca*, and *Schindler's List*. As I look back at the time I was first introduced to Bert the Turtle, in the early 1950s, while attending primary school on the north side of Chicago—America's third largest city and long a favourite hypothetical nuclear target—I realize of course that Bert the

1. ***Duck and Cover.***

Turtle had little to do with culture, history, or aesthetics and much to do with propaganda. America's schoolchildren would never have known what hit them.

The science of nuclear weapons

Atomic energy is the source of power for both nuclear reactors and nuclear weapons. This energy derives from the splitting (fission) or joining (fusion) of atoms. To understand the source of this energy, one must first appreciate the complexities of the atom itself.

An atom is the smallest particle of an element that has the properties characterizing that element. Knowledge about the nature of the atom grew slowly until the early 1900s. One of the first breakthroughs was achieved by Sir Ernest Rutherford in 1911 when he established that the mass of the atom is concentrated in its nucleus; he also proposed that the nucleus has a positive charge and is surrounded by negatively charged electrons. This theory of atomic structure was complemented several years later by Danish physicist Niels Bohr, who placed the electrons in definite shells or quantum levels. Thus an atom is a complex arrangement of negatively charged electrons located in defined shells surrounding a positively charged nucleus. The nucleus, in turn, contains most of the atom's mass and is composed of protons and neutrons (except for common hydrogen, which has only one proton). All atoms are roughly the same size.

Furthermore, the negatively charged electrons follow a random pattern within defined energy shells around the nucleus. Most properties of atoms are based on the number and arrangement of their electrons (Figure 2). One of the two types of particles found in the nucleus is the proton, a positively charged particle. The proton's charge is equal but opposite to the negative charge of the electron. The number of protons in the nucleus of an atom determines what kind of chemical element it is. The neutron is the other type of particle found in the nucleus. Discovered by British physicist Sir James Chadwick, in 1932, the neutron carries no electrical charge and has the same mass as the proton. With a lack of electrical charge, the neutron is not repelled by the cloud of electrons or by the nucleus, making it a useful tool for probing the structure of the atom. Even the individual protons and neutrons have internal structures, called quarks, but these subatomic particles cannot be freed and studied in isolation.

A major characteristic of an atom is its atomic number, which is defined as the number of protons. The chemical properties of an atom are determined by its atomic number. The total number of what are called nucleons (protons and neutrons) in an atom is the atomic mass number. Atoms with the same atomic number but with different numbers of neutrons and, therefore, different atomic masses are called isotopes. Isotopes have identical chemical properties, yet have very different nuclear properties.

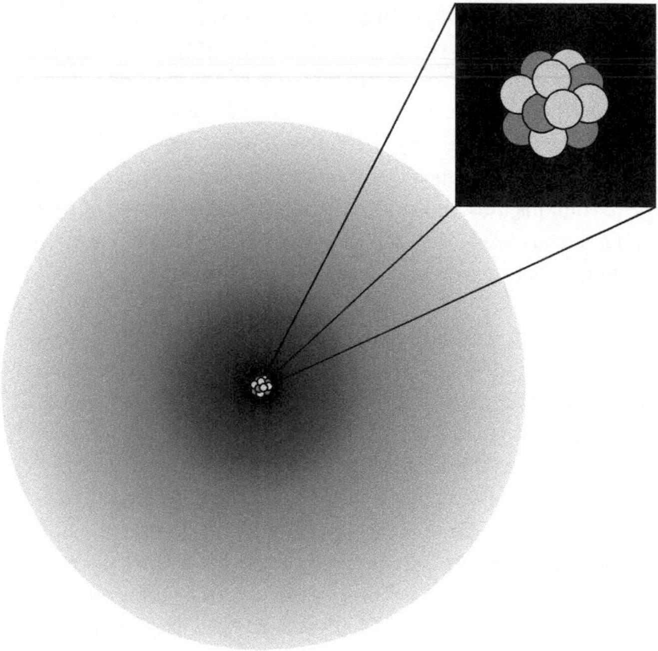

2. Structure of the atom. An atom consists of electrons, protons, and neutrons. The protons and neutrons make up the dense atomic nucleus while the electrons form a more dispersed electron cloud surrounding the nucleus.

For example, there are three isotopes of hydrogen: two of these are stable (not radioactive), but tritium (one proton and two neutrons) is unstable. Most elements have stable isotopes. Radioactive isotopes can also be treated for many elements. The nucleus of the U-235 atom (the chemical sign for uranium is U) comprises 92 protons and 143 neutrons (92 + 143 = 235) and is thus written U^{235}.

The mass of the nucleus is about 1 per cent smaller than the mass of its individual protons and neutrons. This difference is called the *mass defect*, and arises from the energy released when the nucleons (protons and neutrons) bind together to form the nucleus. This energy is called *binding energy*, which in turn determines which nuclei are stable and how much energy is released in a nuclear reaction. Very heavy nuclei and very light nuclei have low binding energies; this implies that a heavy nucleus will release energy when it splits apart (fission) and two light nuclei will release energy when they join (fusion). The mass defect and binding energy are famously related to Albert Einstein's $E = mc^2$.

In 1905 Einstein developed the special theory of relativity, one of the implications of which was that matter and energy are interchangeable with one another. This equation states that a mass (m) can be converted into a tremendous amount of energy (E), where c is the speed of light. Because the speed of light is a large number (186,000 miles a second) and thus c squared is huge, a small amount of matter can be converted into a tremendous amount of energy. Einstein's equation is the key to the power of nuclear weapons and nuclear reactors. Fission reaction was used in the first atomic bomb and is still used in nuclear reactors, while fusion reaction became important in thermonuclear weapons and in nuclear reactor development.

What is the practical significance of a nuclear weapon, then? And how does it differ from what came before? The fundamental difference between a nuclear and conventional weapon is, simply put, that nuclear explosions can be many thousands (or millions) of times more powerful than the largest conventional explosion. To be certain, both types of weapons rely on the destructive force of the blast or shockwave. However, the temperatures reached in a nuclear explosion are very much higher than in a conventional explosion, and a large proportion of the energy in a nuclear explosion is emitted in the form of light and heat, generally referred to as thermal energy. This energy is capable of causing severe skin burns and of starting fires at considerable distances; in fact, damage from the resulting firestorm could be far more devastating than the well-known blast effects.

Nuclear explosions are also accompanied by radioactive fallout, lasting a few seconds, and remaining dangerous over an extended period of time, potentially lasting years. The release of radiation is, in fact, unique to nuclear explosions. Approximately 85 per cent of a nuclear explosion produces air blast (and shock) and thermal energy (heat). The remaining 15 per cent of the energy is released as various types of radiation. Of this, 5 per cent constitutes the initial nuclear radiation, defined as that produced within a minute or so of the explosion, and consisting mostly of powerful gamma rays. The final 10 per cent of the total fission energy represents that of the residual (or delayed) nuclear radiation. This is largely due to the radioactivity of the fission products present in the weapon residues, or debris, and fallout after the explosion.

Equally important is the amount of explosive energy that a nuclear weapon can produce, usually measured as the *yield*. The yield is given in terms of the quantity of conventional explosives or TNT that would generate the same amount of energy when it explodes. Thus, a 1 kiloton nuclear weapon is one that produces the same amount of energy in an explosion as does 1,000 tons of TNT; similarly, a 1 megaton weapon would have the energy equivalent of 1 million tons of TNT.

The uranium-based weapon that destroyed Hiroshima in August 1945, the energy of which resulted from the splitting (fission) of atoms, had the explosive force of 14,000 tons of TNT; the thermonuclear or hydrogen bomb tested by the United States in the Pacific in October 1952, the energy of which came from the joining (fusing) of atoms, had a yield estimated at 7 megatons or 7 million tons of TNT and the production of lethal radioactive fallout from gamma rays. This thermonuclear test was matched by the Soviet Union in August 1953, launching the Cold War superpowers into a deadly race up the nuclear ladder that lasted until the demise of the Soviet Union in December 1991.

Unfortunately, the peaceful end of the Cold War did not mean the end of nuclear threats to global security. Or, to quote former British Prime Minister Tony Blair's defence of his government's plan to update and replace the United Kingdom's Trident nuclear weapons system: 'there is also a new and potentially hazardous threat from states such as North Korea which claims already to have developed nuclear weapons or Iran which is in breach of its non-proliferation duties', not to mention the 'possible connection between some of those states and international terrorism'. Add to this stateless terrorist organizations bent on acquiring the means of mass murder and black market networks of renegade suppliers only too willing to deal in the materials and technical expertise that lead to nuclear weapons, and the picture becomes clearer. The ensuing nightmare of responding to the humanitarian, law and order, and logistical challenges of a nuclear detonation could materialize quite unexpectedly and spectacularly, in any large city, dwarfing the experience of 9/11.

New York City scenario

For example, a relatively small nuclear weapon—say, in the order of a 150 kiloton bomb—constructed by terrorists, detonated in the heart of Manhattan, at the foot of the Empire State Building, at noon on a clear spring day, would have catastrophic consequences. At the end of the first second, the shockwave, causing a sudden change in ambient pressure of 20 pounds per square inch at a distance of four-tenths of a mile from ground zero, would have destroyed the great landmarks of Manhattan, including the Empire State Building, Madison Square Gardens, Penn Central Railroad Station, and the incomparable New York Public Library. Most of the material that constitutes these buildings would remain and pile up to the depth of hundreds of feet in places, but nothing inside this ring would be recognizable. Those caught outside the circle would be exposed to the full effects of the blast, including severe lung and eardrum damage, as well as exposure to flying debris. Those in the direct line of sight of the blast would be exposed to the thermal pulse and killed instantly, while those shielded from some of the blast and thermal effects would be killed as buildings collapse: roughly 75,000 New Yorkers would be killed in these ways. During the next fifteen seconds, the blast and firestorm would extend out for almost four miles, resulting in 750,000 additional fatalities and nearly 900,000 injuries. And this would just be the beginning of New York's problems.

The task of caring for the injured would literally be beyond the ability, and perhaps even the imagination, of the medical system to respond. All but one of Manhattan's large hospitals lie inside the blast area and would be completely destroyed. There aren't enough available hospital beds in all of New York and New Jersey for even the most critically wounded. The entire country has a total of only 3,000 beds in burn centres; thousands would die from lack of medical attention. Meanwhile, most of New York would be without electricity, gas, water, or sewage removal. Transportation of the injured and the ability to bring in necessary supplies, people, and equipment would be problematical. Tens of thousands of New Yorkers would be homeless. The tasks of the emergency responders, in areas that remained dangerously radioactive, would pose possibly insuperable problems.

The terrorists' explosion would have produced much more early radioactive fallout than a similar-sized air burst in which the fireball never touches the ground. This is because a surface explosion produces radioactive particles from the ground as well as from the weapon. The early fallout would drift back to earth on the prevailing wind, creating an elliptical pattern stretching from ground zero out into Long Island. Because the wind would be relatively light, the fallout would be concentrated in the area of Manhattan, just to the east of the blast. Thousands of New Yorkers would suffer serious radiation sickness effects, including chromosomal damage, marrow and intestine destruction, and haemorrhaging. Many would die of these conditions in the days and the weeks ahead. Each survivor of the blast would have on average about a 20 per cent chance of dying of cancer of some form, and another 80 per cent probability of dying instead from other causes such as heart disease or infection.

In January 2007, the scientists who tend the Doomsday Clock moved it two minutes closer to midnight, the ultimate symbol of the annihilation of civilization. The *Bulletin of the Atomic Scientists*, which created the clock in 1947 to warn of the dangers of nuclear weapons, had initially advanced the clock to five minutes to midnight. The Cold War had come and gone, and the danger of nuclear weapons remained front and centre. 'We stand at the brink of a second nuclear age', the group said in a statement sixty year later, pointing to North Korea's first test of a nuclear weapon in 2006, Iran's nuclear ambitions, US flirtation with atomic 'bunker busters', and the thousands of operational nuclear weapons then available to the nuclear club and potential terrorists. The scientists also reminded us that only fifty of today's nuclear weapons could kill as many as 200 million people. In 2019, in a sign of a new abnormal, which also includes the threat of climate change, the clock is *still* two minutes to midnight. Scientists now had to contend with the growing fragility of the existing arms-control regime and to possible consequences of its gradual disappearance.

Since it was set to seven minutes to midnight in 1947, the hand of the Doomsday Clock has moved eighteen times. It came closest to midnight—two minutes away—not surprisingly, in early 1953, following the successful test of America's hydrogen bomb, code-named 'Mike', which somehow managed to vaporize the Pacific island test site. This was about the same time that I was first introduced to Bert the Turtle and his sombre warning, 'duck and cover'.

Chapter 2
Building the bomb

Since late 1944, American long-range B-29 bombers had been conducting the greatest air offensive in history. In total, approximately 160,000 tons of bombs were dropped upon Japan towards the end of the war, including fire-bomb raids that destroyed downtown Tokyo and a number of other large Japanese cities. These raids alone killed 333,000 Japanese soldiers and civilians and wounded half a million more.

Massive loss of life and property in this manner was not unprecedented. Up until the Nazi surrender in May 1945, 635,000 Germans, mostly civilians, died and 7.5 million were made homeless when British and US bombs were dropped on 131 cities and towns. The rationale was simple enough. 'The idea is', observes German revisionist Jorg Friedrich, in his study of the Allied bombing of Germany during World War II, 'that the cities and their production and their morale contributed to warfare. So warfare is not simply the business of an army, it's the business of a nation.' In total war, everything and everyone becomes a target. This of course was not news to contemporaries such as George Orwell, who reminds us in the great essay 'England Your England', written in February 1941, with the Luftwaffe overhead: 'highly civilized beings are flying overhead, trying to kill me'.

It was now the turn of Hitler's allies. The Japanese war economy was all but destroyed. But still Japan refused to surrender. Although there were elements within the Japanese government that had long recognized that the war was lost, official Allied policy continued to be nothing less than unconditional surrender. So, while Japanese civilian leaders and Emperor Hirohito favoured suing for peace, the militarists, led by the army, resisted. Faced with such determined resistance, the US Chiefs of Staff estimated that the human costs of invading the Japanese home islands would be no fewer than one million US and Allied casualties. Deeply troubled by such a prospect, President Harry S. Truman, who had succeeded to the presidency after the sudden death of Franklin D. Roosevelt on 12 April 1945, sought alternatives.

For his part, Secretary of War Henry L. Stimson eagerly instructed President Truman on the implications of the potentially devastating new weapon being developed at the top-secret Manhattan Project. On 23 April, Stimson and General Leslie Groves, the project director, gave the new president a lengthy briefing on the weapon we now know as the atomic bomb. Here Groves reported on the genesis and current status of the atomic bomb project, while Stimson presented a memorandum explaining the implication of the bomb for international relations. Stimson addressed the terrifying power of the new weapon, advising that 'within four months, we shall in all probability have completed the most terrible weapon ever known in human history, one bomb which could destroy a whole city'. He went on to allude to the dangers that its discovery and development foreshadowed, and pointed to the difficulty in constructing a realistic system of controls.

Truman seemed to focus less on the geopolitical implications of the possession of the atomic bomb and more on the personal burden of having to authorize the use of the awesome weapon.

'I am going to have to make a decision which no man in history has ever had to make', he reportedly said to a White House staffer, the very next person he saw after Stimson and Groves left his office. 'I will make the decision, but it is terrifying to think about what I will have to decide.' In time, Truman would make a choice, probably with insufficient forethought, based on his own wartime experience and information at hand.

Origins of the Manhattan Project

Though no single decision created the American atomic bomb project, most accounts begin with the presidential discussion of a letter written by the most famous scientist of the 20th century, Albert Einstein. On 11 October 1939, Alexander Sachs, Wall Street economist and unofficial adviser to President Roosevelt, met with the president to discuss a letter written by Einstein on 2 August. Einstein had written to inform Roosevelt that recent research had made it 'probable . . . that it may become possible to set up a nuclear chain reaction in a large mass of uranium, by which vast amounts of power and large quantities of new radium-like elements could be generated', leading 'to the construction of bombs, and it is conceivable—though much less certain—that extremely powerful bombs of a new type may thus be constructed'. This was all likely to happen 'in the immediate future'.

Einstein believed, rightly, that the Nazi government was actively supporting research in the area and urged the US government to do the same. Sachs read from a cover letter he had prepared and briefed Roosevelt on the main points contained in Einstein's letter. Initially the president was noncommittal and expressed concern over the necessary funds, but at a second meeting over breakfast the next morning Roosevelt became persuaded of the value of exploring atomic energy. He could hardly do otherwise.

Einstein drafted his famous letter with the help of Hungarian émigré Leó Szilárd, one of a number of brilliant European physicists who had fled to America in the 1930s to escape Nazi and fascist repression. Szilárd was among the most vocal of those advocating a programme to develop bombs based on recent findings in nuclear physics and chemistry. Those like Szilárd, and fellow Hungarian refugee physicists Edward Teller and Eugene Wigner, regarded it as their ethical responsibility to alert America to the possibility that German scientists might win the race to build an atomic bomb and to warn that Hitler would be more than willing to resort to such a weapon. But Roosevelt, preoccupied with events in Europe, took over two months to meet with Sachs after receiving Einstein's warning. Szilárd and his colleagues had initially interpreted Roosevelt's apparent inaction as unwelcome evidence that the Americans did not take the threat of nuclear warfare seriously. They were wrong.

Roosevelt wrote back to Einstein on 19 October 1939, informing the physicist that he had set up an exploratory committee consisting of Sachs and representatives of the army and navy to study uranium. Events proved that the president was a man of considerable action once he had chosen a course of direction. In fact, Roosevelt's approval of uranium research in October 1939, based on his belief that the United States could not take the risk of allowing Hitler to achieve unilateral possession of 'extremely powerful Bombs', was the first of many decisions that ultimately led to the establishment of the only atomic bomb effort that succeeded in World War II.

By the beginning of World War II, there was growing concern among scientists in the Allied nations that Nazi Germany might be well on its way to developing fission-based weapons. Organized research first began in Britain as part of the Tube Alloys project, and in America a small amount of funding was given for research into uranium weapons, starting in 1939 with the Uranium Committee under the direction of Lyman J. Briggs. At the urging of British scientists though, who made crucial calculations indicating that a fission weapon could be completed in only a few years, by 1941 the project had been wrestled into better bureaucratic hands, and in 1942 came under the auspices of the Manhattan Project. The project brought together the top scientific minds of the day, including many exiles from Nazi Europe, with the production power of American industry, for the single purpose of producing fission-based explosive devices before the Germans. London and Washington agreed to pool their resources and information, but the other Allied partner—the Soviet Union under Joseph Stalin—was not informed.

Berlin, Tokyo, and the bomb

The Allied scientists had much to fear from Berlin. Late in 1938, Lise Meitner, Otto Hahn, and Fritz Strassman discovered the phenomenon of atomic fission. Meitner worked in Germany with physicists Hahn and Strassman until fleeing to Sweden to escape Nazi persecution. From her work in Germany, Meitner knew the nucleus of uranium-235 splits (fission) into two lighter nuclei when bombarded by a neutron, and that the sum of the particles derived from fission is not equal in mass to the original nucleus. Moreover, Meitner speculated that the release of energy—energy a hundred million times greater than normally released in the chemical reaction between two atoms—accounted for the difference. In January 1939, her nephew, the physicist Otto Frisch, substantiated these results and, together with Meitner, calculated the unprecedented amount of energy released. Frisch applied the term 'fission', from biological cell division, to name the process. Danish physicist Niels Bohr sailed for the US shortly thereafter and announced the discovery. In August, Bohr and John A. Wheeler, working at Princeton University, published their theory that the isotope uranium-235, present in trace quantities within uranium-238, was more fissile than uranium-238 and should become the focus of uranium research. They also postulated that a then unnamed, unobserved transuranic element, aptly referred to as 'high octane', produced during the fissioning of uranium-238, would be highly fissionable. Enrico Fermi and Leó Szilárd quickly realized that the first split or fission could cause a second, and so on in a series of chain reactions, expanding in geometric progression. This was the moment Szilárd and fellow atomic scientists persuaded Einstein to write to Roosevelt.

Physicists everywhere soon recognized that if the chain reaction could be tamed, fission could lead to a promising new source of power. What was needed was a substance that could 'moderate' the energy of neutrons emitted in radioactive decay, so that they could be captured by other fissionable nuclei, with heavy water a prime candidate for the job. After the discovery of fission, German Nobel Prize laureate Werner Heisenberg was recruited to work on a chain-reacting pile in September 1939 by Nazi physicist Kurt Diebner. While the Americans under Fermi chose graphite to slow down or moderate the neutrons produced by the fission in uranium-235 so that they could cause further fissions in a chain reaction, Heisenberg chose heavy water.

Heisenberg calculated the critical mass for a bomb in a 6 December 1939 report for the German Arms Weapons Department. His formula, with the nuclear parameters value assumed at that time, yielded a critical mass in the hundreds of tons of 'nearly' pure uranium-235 required for an exploding reaction, Heisenberg's model for a bomb at the time. This was vastly beyond what Germany could hope to produce. With uranium out of the question, the Germans opted for plutonium, which meant building an atomic pile or nuclear reactor to convert natural uranium into plutonium. Unlike America's Manhattan Project, the Nazi nuclear physics programme was never able to produce a critical nuclear reactor, despite the efforts of Heisenberg and Diebner. The Nazi attempt to build a reactor, in fact, proved feeble and disorganized, while their effort to build an atomic weapon was non-existent. But the Allies did not know that. Nor did they know much about Japan's efforts to create a nuclear weapon.

In Tokyo, in autumn 1940, the Japanese army concluded that constructing an atomic bomb was indeed feasible. The Institute of Physical and Chemical Research, or Rikken, was assigned the project under the direction of Yoshio Nishina. The Imperial Navy was also diligently working to create its own 'superbomb' under a project dubbed F-Go (or No. F, for fission), headed by Bunsaku Arakatsu, towards the end of 1945. The F-Go programme had begun life in Kyoto in 1942. However, the military commitment wasn't backed with adequate resources, and the Japanese effort to build an atomic bomb had made little progress by the end of the war.

Japan's nuclear efforts were disrupted in April 1945 when a B-29 raid damaged Nishina's thermal diffusion separation apparatus. Some reports claim the Japanese subsequently moved their atomic operations to Hungnam, now part of North Korea. The Japanese may have used this facility for making small quantities of heavy water. The Japanese plant was captured by Soviet troops at the war's end, and some reports claim that the output of the Hungnam plant was collected every other month by Soviet submarines, as part of Moscow's own nuclear energy programme (see Chapter 4).

There are indications that Japan had a more sizeable programme than is commonly understood, and that there was close cooperation among the Axis powers, including the secretive exchange of war materiel. The Nazi submarine U-234, which surrendered to American forces in May 1945, was found to be carrying 560 kilograms of uranium oxide destined for Japan's own atomic programme. The oxide contained about 3.5 kilograms of the isotope U-235, which would have been one-fifth of the total U-235 needed to make one bomb. After Japan surrendered in August 1945, the occupying US army found five Japanese cyclotrons, which could be used to separate fissional material from ordinary uranium. The Americans smashed the cyclotrons and dumped them into Tokyo harbour.

The road to Trinity

A massive industrial and scientific undertaking, employing 65,000 workers, the Manhattan Project involved many of the world's great physicists in the scientific and development aspects. For its part, the United States made an unprecedented investment into wartime research for the project, which was spread over thirty sites in the US and Canada. The actual design and construction of the weapon was centralized at a secret laboratory in Los Alamos, New Mexico, previously a small ranch school near Santa Fe. The laboratory that designed and fabricated the first atomic bombs began to take shape in spring 1942 with the recommendation that the US Office of Scientific and Research Development and the army look at ways to further bomb development. By the time of his appointment in late September, General Groves had orders to set up a committee to study military applications of the bomb. Shortly thereafter, J. Robert Oppenheimer headed the work of a group of theoretical physicists he called the luminaries, which included Felix Bloch, Hans Bethe, Edward Teller, and Robert Seber, while John H. Manley assisted him by coordinating nationwide fission research and instrument and measurement studies from the Metallurgical Laboratory in Chicago. Despite inconsistent experimental results, the consensus emerging at Berkeley (from where most of the scientists had been seconded) was that approximately twice as much fissionable material would be required than had been estimated six months earlier. This was disturbing, especially in light of the military's view that it would take more than one bomb to win the war.

In many ways, the Manhattan Project operated like any other large construction company. It purchased and prepared sites, let contracts, hired personnel and subcontractors, built and maintained housing and service facilities, placed orders for materials, developed administrative and accounting procedures, and established communications networks. By the end of the war, General Groves and his staff had spent approximately $2.2 billion on, among other things, production facilities and towns in the states of Tennessee, Washington, and New Mexico, as well as on research in university laboratories from Columbia University, in New York City, to the University of California at Berkeley. What made the Manhattan Project clearly unlike other companies performing similar functions was that, because of the necessity of moving quickly, it invested hundreds of millions of dollars in unproven and hitherto unknown processes, and did so entirely in secret. Speed and secrecy were the watchwords of the Manhattan Project.

Secrecy proved to be a blessing in disguise. Although it dictated remote site locations, required subterfuge in obtaining labour and supplies, and served as a constant irritant to the academic scientists on the project, it had one overwhelming advantage: secrecy made it possible to make decisions with little regard for normal peacetime considerations. Groves knew that as long as he had the backing of the president, money would be available and he could devote his energies entirely to the running of the project. Secrecy was so complete that many of the staff did not know what they were working on until they heard about the bombing of Hiroshima on the radio.

Moreover, the need for haste clarified priorities and shaped decision making. Unfinished research on three separate, unproven processes had to be used to freeze design plans for production facilities, even though it was recognized that later findings would dictate changes. The pilot stage was eliminated entirely, violating all manufacturing practices and leading to intermittent shutdowns and endless troubleshooting during trial runs in production facilities. The inherent problems of collapsing the stages between the laboratory and full production created an emotionally charged atmosphere, with optimism and despair alternating with confusing frequency.

Despite Groves' assertion that an atomic bomb could probably be produced by 1945, he and other principals associated with the project fully recognized the magnitude of the tasks before them. For any large organization to take laboratory research into design, construction, operation, and product delivery in two and a half years (from 1943 to August 1945) would have been a major industrial achievement. Whether the Manhattan Project would be able to produce bombs in time to affect the outcome of World War II was an altogether different question as 1943 began. And, obvious though it seems in retrospect, it must be remembered that no one at the time knew the war would end in 1945 or, equally important, who the remaining adversaries would be when and if the atomic bomb was ready to use.

At precisely 5.30 a.m., on Monday 16 July 1945, at 'Trinity', the code name for the Manhattan Project test site in the desert south-east of Socorro, New Mexico, a group of officials and scientists led by Groves and Oppenheimer witnessed the first explosion of an atomic bomb. And what a show it was. A pinprick of brilliant light punctured the darkness of the New Mexico desert, vaporizing the tower and turning asphalt around the base of the tower to green sand. The bomb released the explosive force of nearly 21,000 tons of TNT, and the New Mexico sky was suddenly brighter than many suns. Some observers suffered temporary blindness even though they looked at the brilliant light through smoked glass. Seconds after the explosion came a huge blast, sending searing heat across the desert and knocking some observers, standing 1,000 yards away, to the ground. A steel container weighing over 200 tons, standing half a mile from ground zero, was knocked ajar. As the orange and yellow fireball stretched up and spread, a second column, narrower than the first, rose and flattened into a mushroom cloud, providing the atomic age with a symbol that has since become imprinted on the human consciousness. *New York Times* reporter William Laurence called the explosion 'the first cry of a new-born world'.

For a fraction of a second, the light produced by Trinity was greater than any ever before produced on earth, and could have been seen from another planet. And as the light dimmed and the mushroom cloud rose, Oppenheimer was reminded of fragments from the *Bhagavad-Gita*, the sacred Hindu text: 'I am become Death | The destroyer of worlds'. Less quoted but more memorable perhaps was the comment by test site manager Kenneth Bainbridge to Oppenheimer: 'Oppie, now we're all sons of bitches.' The terrifying destructive power of atomic weapons and the uses to which they could be put were to haunt many of the Manhattan Project scientists for the remainder of their lives.

By the end of July, the Manhattan Project had produced two different types of atomic bomb, code-named 'Fat Man' and 'Little Boy' (Figures 3 and 4). Fat Man was the more complex of the two. A bulbous, ten-foot bomb containing a sphere of metal plutonium-239, it was surrounded by blocks of high explosives that were designed to produce a highly accurate and symmetrical implosion. This would compress the plutonium sphere to a critical density and set off a nuclear chain reaction. Scientists at Los Alamos were not altogether confident in the plutonium bomb design—hence the necessity of the Trinity test. The Little Boy type of bomb had a much simpler design than Fat Man. Little Boy triggered a nuclear explosion, rather than implosion, by firing one piece of uranium-235 into another.

When enough U-235 is brought together, the resulting fission chain reaction can produce a nuclear explosion. But the critical mass must be assembled very quickly, otherwise the heat released at the start of the reaction will blow the fuel apart before most of it is consumed. To prevent this inefficient pre-detonation, the uranium bomb used a gun to fire one piece of U-235 down the barrel into another. Moreover, the bomb's gun-barrel shape was believed to be so reliable that testing was ruled out. Interestingly, testing would have been out of the question anyway, since producing Little Boy had used all the purified U-235 produced to date. Clearly, though, the Manhattan Project had managed to take the discovery of fission from the laboratory to the battlefield.

3. A replica of the 'Fat Man' bomb.

4. A replica of the 'Little Boy' bomb.

The Hiroshima decision

General Groves quickly conveyed word of the test to Secretary of War Stimson's aide, who in turn relayed word to his boss in cryptic fashion: 'Operated on this morning. Diagnosis not yet complete but results seem satisfactory and already exceed expectations.' Stimson, filled with excitement, gave Truman a preliminary report in the evening, after the president returned from his tour of Berlin while still at the Potsdam Conference. While the success of the bomb took a great load off his mind, Truman, up to then uncertain whether he would need Soviet assistance to finish off the Japanese, casually informed Stalin that the US 'had a new weapon of unusual destructive force'. Stalin, who had spies on the ground in New Mexico, simply replied that he hoped he would use it well. Certainly, with the success of 'Trinity', the US government believed that it could probably conclude the war without Russian assistance, and from Potsdam, Truman sent an ultimatum to Tokyo to surrender immediately, unconditionally, or face 'prompt and utter destruction'.

In any case, the US now had in its arsenal a weapon of unparalleled destruction; Stimson even suggested that it would create 'a new relationship of man to the universe'. Truman's advisers agreed that the atomic bomb could end the war in the Pacific, but they could not agree on the best way to use it. There is a certain irony here: the scientists who developed the bomb wanted it used against the Nazis and were horrified when it became clear it would be used against Japan. Some proposed a public demonstration on an uninhabited region; others argued that it should be used against Japanese naval forces and should never be used against Japanese cities. Still others argued that the objective was not so much to defeat Japan as to employ 'atomic diplomacy' against the Soviet Union, providing a demonstration to make it more manageable in Eastern and Central Europe after the war.

After considering the various proposals, Truman concluded that the only way to shorten the war, while avoiding an invasion of Japan, was to use the bomb against Japanese cities. On the morning of 6 August 1945, shortly after 8.15 a.m., a lone B-29 bomber named the *Enola Gay* dropped Little Boy over the city of Hiroshima (population 350,000), Japan's second most important military-industrial centre, instantly killing 80,000 to 140,000 people and seriously injuring 100,000 or more (Figure 5). The first (never-before tested) uranium-235-based bomb to be used had the explosive force of 14,000 tons of TNT—puny and primitive by later thermonuclear standards. Still, in that one terrible moment 60 per cent of Hiroshima, four square miles, an area equal to one-eighth of New York City, was destroyed. The burst temperature was estimated to reach over a million degrees Celsius, which ignited the surrounding air, forming a fireball some 840 feet in diameter. Eyewitnesses more than five miles away said its brightness exceeded the sun tenfold. The blast wave shattered windows for a distance of ten miles and was felt as far away as thirty-seven miles. Over two-thirds of Hiroshima's buildings were demolished. The hundreds of fires, ignited by the thermal pulse, combined to produce a firestorm that incinerated everything within about 4.4 miles of ground zero. Hiroshima had disappeared under a thick, churning foam of flame and smoke.

Three days later, on 9 August, another lone B-29 bomber, named *Bock's Car*, dropped Fat Man (the Trinity test bomb) on Nagasaki (population 253,000), home to two huge Mitsubishi war plants on the Urakami River, instantly killing 24,000 and wounding 23,000. The plutonium bomb had the explosive force of 22,000 tons of TNT, more than 2,000 times the blast power of what had previously been the world's most devastating bomb, the British 'Grand Slam', a logical technological improvement in the strategy of city-busting that the Allies had developed at Hamburg and Dresden. But unlike Hiroshima, there was no firestorm this time. Despite this, the blast was more destructive to the immediate area, due to the topography and the greater power of Fat Man. However, the hilly, almost

5. Atomic bomb cloud over Hiroshima.

mountainous terrain limited the total area of destruction to less than that of Hiroshima, and the resulting loss of life was also not as great. With Japanese doctors at a loss to explain why many civilian patients who had not been wounded were now wasting away, in the following weeks the death counts in both cities rose as the populations succumbed to radiation-related sickness.

The shockwaves were felt well beyond the Japanese home islands. Western newspapers struggled to explain to a triumphant but mystified public how thousands of American, British, and Canadian scientists had managed to harness the power of the sun to such deadly effect. No easier to explain was that the US government could undertake a military and scientific programme as massive and prolonged as the Manhattan Project in such absolute secrecy. This paradoxical view of the government's achievement was typical of the American public's response to the bomb. The elation at the prospect of imminent peace was tempered by a growing recognition of the awesome responsibilities of possessing such a powerful weapon. Critics such as British scientist P. M. S. Blackett argued that Hiroshima and Nagasaki could best be seen as the first chapter of the Cold War rather than the last chapter of World War II. Opposition to nuclear energy emerged almost immediately after the bomb was built. The Franck Report of 11 June 1945, signed by a number of the Manhattan Project scientists, warned Secretary of War Stimson that an unannounced attack would surely lead to an arms race. Both the report and the scientists were ignored.

The impact of the new weapon spread well beyond the military and scientific circles in which it had been developed; to an extent unprecedented, it began to seep into the popular imagination as images of mushroom clouds became symbolic of the new destructive potential that had been created. What Truman called 'the greatest scientific gamble in history' had paid off with devastating effectiveness, and there was no doubt that a turning point in the history of the contemporary world had been reached. Indeed, 'the bomb', as it was quickly dubbed, became the defining feature of the post-World War II world.

With a Japanese surrender imminent, and recognizing that if it was going to play a part in post-war Asia it would need to enter the fray quickly, the Soviet Union declared war on Japan on

8 August, a week sooner than Stalin had pledged at the Potsdam Conference. Nine minutes after its declaration, the Soviet Union's Far Eastern army and air force launched a massive offensive against the Japanese forces in Manchuria and the Korean peninsula. The seizure of the Kurile Islands and southern Sakhalin also constituted part of the Soviet continental campaign. The overwhelming nature of the Soviet attack caused very high casualties among the Kwantung army, killing 80,000 Japanese soldiers (against 8,219 Soviet dead and 22,264 wounded) in less than a week. The writing was on the wall.

Yielding to the reality of the situation, Emperor Hirohito, supported by civilian advisers, finally overcame the xenophobic fanaticism of the militarists, who reasoned that they had been defeated through superior science rather than by valour and force of arms, and ordered surrender on 14 August. For its part, the United States agreed to retain the institution of the emperor system, stripped of pretension to divinity and subject to American occupation headed by General Douglas MacArthur. On 2 September, thereafter known as V-J Day, a great Allied fleet sailed into Tokyo Bay. Aboard the *USS Missouri*, General MacArthur accepted Japanese surrender on behalf of the Allies. With this simple ceremony, World War II was brought to a close.

President Truman, who ordered the bombings, insisted that his decision had shortened the war and prevented huge casualties. The historical evidence strongly suggests that he was right. Any US president probably would have acted similarly, as a war-weary nation faced the prospect of further fighting and appalling losses. For the victors, the bombings of Hiroshima, and three days later Nagasaki, were terrible acts of war. But they were no crime. The vanquished, with few exceptions, would choose to differ.

Chapter 3

A choice between the quick and the dead

When we contemplate the origins and issues of nuclear disarmament in the immediate aftermath of World War II, we should bear in mind that at the beginning of the nuclear age there were no rules, no non-proliferation norms, no concept of nuclear deterrence, and, particularly, no taboo against nuclear war. There was, however, an apparent arms race, hard on the heels of a conflict that probably killed sixty million souls. At the same time, advances in atomic energy held out prospects for important peaceful uses, such as nuclear power providing limitless energy to the world. Significantly, the processes associated with the military and civilian uses of atomic energy were virtually the same.

Traditionally, as with most scientific advances, there were efforts to share information at the international level. But because of the well-known destructive ability of the atomic bomb and the power that it gave its possessor, America was in no mood to share its nuclear secrets in the absence of an effective international control system. Reconciling the drive to reap the peaceful benefits of this newly harnessed force with the need to control its destructive potential was always going to pose a problem.

Early efforts focused on countering the problem with international agreements and tied non-proliferation with disarmament. Not two months had passed since Hiroshima when

President Harry S. Truman told Congress that: 'The hope of civilization lies in international arrangements looking, if possible, to the renunciation of the use and development of the atomic bomb.' It was a view widely shared by influential atomic scientists. The Franck Report, named after the chairman of the committee issuing the report, recommended in June 1945, before the atomic bombs were dropped on Japan, that since a perpetual US monopoly would likely be impossible to maintain, the elimination of nuclear weapons would have to be realized through international agreements.

Several political actions occurred that were aimed at establishing a framework in which to consider the control of atomic energy. The Three Nation Agreed Declaration was concluded among the United States, Great Britain, and Canada, wartime partners in the development of the bomb. On 15 November 1945, in Washington, the three countries declared their intent to share with all nations the scientific information associated with atomic energy for peaceful or civilian purposes. Recognizing the dilemma of reconciling the peaceful and destructive powers of atomic energy, the declaration called for withholding this information until appropriate safeguards were in place. It then called on the United Nations to establish a commission to recommend a system of international control.

At the Conference of Ministers meeting, held in Moscow on 27 December 1945, the Soviet Union agreed to these principles in the Moscow Declaration, a Soviet-Anglo-US statement. The declaration also contained the text of a proposed UN resolution to establish a commission on controlling atomic energy; it invited France, China, and Canada to co-sponsor the resolution, which was passed unanimously during the first session of the UN General Assembly, on 24 January 1946.

In this way, the United Nations Atomic Energy Commission (UNAEC) was established. It consisted of all members of the

UN Security Council (Australia, Brazil, China, Egypt, France, Mexico, the Netherlands, Poland, the Soviet Union, the United Kingdom, and the United States), together with Canada—a total of twelve countries. The resolution called for the commission to be accountable to the Security Council, dominated by America, Britain, China, and the Soviet Union. This move, suggested by Moscow, demonstrated how the efforts to share atomic knowledge would be dominated by Security Council considerations. The Security Council also operated with a veto power for each permanent member on substantive but not procedural issues. The veto—then and now—would play an important role in the efforts to control the atom.

The responsibilities of the UNAEC included: overseeing the exchange of basic scientific information for peaceful ends; control of atomic energy to ensure its use for only peaceful ends; the elimination of atomic weapons from national arsenals; and effective safeguards by way of inspection and other means to protect complying states against the hazards of violation and evasion.

At the same time, Secretary of State James F. Byrnes formed a committee to study methods of control and safeguards to protect the United States during the negotiations. The five members of the group, led by Assistant Secretary of State Dean Acheson, were drawn from military and political circles associated with the bomb's development. Acheson's committee looked to a 'Board of Consultants' as a source of knowledge on technical aspects of atomic energy. The board was led by David Lilienthal, chairman of the Tennessee Valley Authority, and included three other scientists, notably J. Robert Oppenheimer, the physicist who played a major role in the Manhattan Project.

The combined effort of these two groups resulted in a document entitled 'A Report on the International Control of Atomic Energy', which promptly became known as the 'Acheson–Lilienthal Report'. Released in late March 1946, the report highlighted the technical characteristics that would determine the nature of an international control system. More importantly, the conferees regarded their conclusions as a foundation for discussion rather than a final plan. The United States' proposal at the UNAEC would draw heavily on the Acheson–Lilienthal Report's ideas for a system of international control.

The Baruch Plan

This is the background behind the US proposals made to the United Nations in June 1946. Known as the Baruch Plan after its chief negotiator Bernard Baruch, the elder statesman who had served American presidents in various capacities since World War I, the plan's objective was to prevent the further spread of nuclear weapons, ostensibly by securing atomic technology and materials through the control of the newly created United Nations. Under the plan, a UN authority would supervise and control the mines of the raw materials for atomic weapons and would be responsible for any production. Also, under the plan, the United States would give up its atomic weapons and facilities in a phased transition.

In presenting the plan to the United Nations on 14 June 1946, Baruch employed a melodramatic allusion to America's Wild West past: 'We are here to make a choice between the quick and the Dead . . . If we fail, then we have damned every man to be a slave of Fear. Let us not deceive ourselves: We must elect World Peace or World Destruction.' The fundamentals of the Baruch Plan were easy enough for the public to grasp. The former chairman of Woodrow Wilson's War Industries Board proposed the creation of an International Atomic Development Authority whose sole duty would be to oversee all phases of the development and use of atomic energy; the key to the successful operation of such an agency would be its effectiveness in controlling and inspecting atomic energy activities—for then, and only then, would the United States be prepared both to cease the manufacture of atomic weapons and dispose of its stockpile.

Baruch listed several activities that would be regarded as criminal: possession or separation of atomic material suitable for use in a bomb; seizure of property owned or licensed by the authority; interference with the authority's activities; and engaging in 'dangerous' projects that were contrary to, or without a licence from, the authority. Then, making his own distinctive contribution, Baruch called for severe penalties to be imposed on countries that engaged in such activities. And although he conceded the importance of the veto power to the work of the Security Council, he said that with respect to atomic energy, 'there must be no veto to protect those who violate their solemn agreement not to develop or use atomic energy for destructive purposes'.

Responses to the plan varied widely. After reading the speech, Winston Churchill praised Baruch, saying 'There is no man in whose hands I would rather see these awful problems placed than Bernard Baruch's.' Some opposed it for giving too much away; others opposed it as being unfair to the Soviets and called for an immediate halt in the manufacture of atomic bombs. Some thirty senators said the plan was not tenable, while Senate Foreign Relations Committee Chairman Arthur Vandenberg said the plan was 'more important to the peace of the world than anything that happened in New York'. By September, one survey reported that 78 per cent of the American public endorsed the plan.

The issue of the veto prompted both favourable and critical comments. The famous columnist Walter Lippmann accused Baruch of taking America up a blind alley with the veto provision, while Supreme Court Justice William O. Douglas supported Baruch's proposal to strip the Security Council of its veto in atomic matters. The American *Daily Worker*, the US Communist Party paper, saw the elimination of the veto as an opportunity for Washington and London to 'carry the day' against the Soviet Union, 'demonstrating a new predatory flight of the American eagle'. The Kremlin's response to the Baruch Plan came five days later, on 19 June, in an address delivered by Soviet Deputy Foreign Minister Andrei Gromyko.

The Gromyko Plan

Sidestepping the American case for atomic peace, Gromyko instead called for an international convention aimed at prohibiting the production and employment of atomic weapons, while demanding unilateral nuclear disarmament by the United States as a precondition for any agreement. To this end, he introduced two resolutions. The first called for the convention to ban the use and production of atomic bombs, destroying existing weapons within three months, while calling on the principals to pass laws within their own countries to punish violators; the second called for the formation of two committees, one for exchanging scientific information and the other to find ways to ensure compliance with the provisions.

The only direct response to the Baruch Plan came in the form of Soviet opposition to eliminating the veto: 'Attempts to undermine the principles, as established by the Charter of the Security Council, including the unanimity of the members of the security council in deciding questions of substance, are incompatible with the interests of the United Nations . . . [and] must be rejected.' It was unlikely that Joseph Stalin's representative could have said anything else, as the Cold War lines were being drawn.

Washington's official reaction was low key. In a press conference, one member of the US delegation said he was not discouraged and characterized the Soviet proposal 'by way of argument rather than a final Soviet position'. In order to avoid an open split at this early stage of negotiations, the American delegation used anonymous stories in the press to make its point. Accordingly, the *New York Times* reported that according to a reliable source, the United States was not able to accept the Gromyko Plan, at least not in the absence of the safeguards proposed by Baruch, as it meant giving up America's source of military power.

Initially, the UNAEC agreed to break up into a working committee of the whole, in order to draft a plan incorporating all of the ideas suggested for the international control authority. Both Washington, noting the level of support for its proposal, and Moscow reiterated their respective positions. After some delay with Gromyko over its name, a smaller group—Subcommittee One—was formed to draft possible features that a control plan might have; the membership of Subcommittee One was composed of France, Mexico, Britain, the United States, and the Soviet Union.

Subcommittee One met on 1 July, a day after the United States had conducted a test of an atomic bomb at Bikini Atoll, evidence to some that America had no intention of relinquishing its monopoly over the bomb. In addition to handing a propaganda victory to the Soviets, continuing the US tests may have provided the impetus for them to pursue their own. Another test was held on 25 July. In September, however, Truman postponed the next test—scheduled for March 1947—partly out of deference to the negotiations.

The discussions in Subcommittee One highlighted some of the basic differences between the sides. Gromyko insisted on outlawing atomic weapons first, and was less concerned about a system of control. For their part, the Americans demanded adequate control before they would give up their weapons.

The opposing positions of each country on the veto issue became further entrenched. And although the goal of the Americans in submitting the memoranda had been to elicit more specific responses from the Soviets, Gromyko held his ground.

The chairman of Subcommittee One, Australian external affairs minister Herbert Evatt, recognized the impasse and proposed to the full UNAEC that three committees of the whole be formed to address technical questions, leaving political questions aside, all in the hope of finding common ground. By majority vote, the group formed Committee Two, the Scientific and Technical Committee (the only one whose formation the Soviets supported), and a Legal Committee. The most important work occurred in the Scientific and Technical Committee.

Committee Two met first, but was unable to move beyond the differences experienced in Subcommittee One and became the forum for Gromyko's outright rejection of the Baruch Plan. In sum, he said, on 24 July 1946, 'The United States' proposals in their present form cannot be accepted in any way by the Soviet Union either as a whole or in separate parts.' He also refused to yield to the elimination of the veto. Harking back to the founding of the United Nations, Gromyko underscored the importance of the issue of sovereignty in the deliberations. He addressed the Baruch Plan to consider atomic energy as a matter of international and not of national importance. Accordingly, he viewed this principle as a violation of Article 2, Paragraph 7 of the United Nations Charter, which called for no interference in the internal affairs of member states.

The Scientific and Technical Committee had begun meeting on 19 July 1946, and the framework within which the members operated proved highly successful. Forming an informal group of scientists, the committee agreed that no one in the group would represent his country; the members would simply explore the technical aspects of safeguards as individuals. Whatever conclusions they drew would be referred back to the main committee. The United States provided, in addition to the technical information in the Acheson–Lilienthal Report, background information and information on the beneficial use of atomic energy in eleven different treatises. In response to its mandate, the committee completed its report on 3 September, concluding that it had been unable to find 'any basis in the available scientific facts for

supposing that effective control is not technologically feasible'.
There was always going to be another problem—the political kind.

As the work of the committee became bogged down, Baruch decided to write a letter to Truman seeking approval for two recommendations. The first would be to force a vote in the UNAEC at an early date, preferably before January 1947, when the membership of the commission would rotate; the second would be a call for military preparedness in the field of atomic energy, in the likely failure of the UNAEC.

Widespread news coverage of the views of Secretary of Commerce Henry Wallace, who was scathing of the Baruch Plan, provided the backdrop of Baruch's visit to the White House to deliver his letter on 18 September. Wallace's remarks, well received by a liberal audience, had hit Baruch to the quick, badly undercutting him publicly. Wallace said that a major defect of the Baruch Plan was America's insistence that other countries give up their right to explore military uses of nuclear energy and turn over raw materials to an international authority, whereas the United States would not give up its weapons until it deemed such a system was in place. Wallace did not believe the US would be amenable to such a deal if the tables were turned.

To Baruch, such a display of disunity could only undermine the impact of the coming UNAEC vote. At the Paris Peace Conference of Foreign Ministers, Secretary of State James F. Byrnes made a similar complaint, arguing that Wallace's statements had eroded his own position there. Both Baruch and Byrnes threatened to resign if Wallace did not recant. With the writing on the wall, Truman asked for, and received, Wallace's resignation, on 20 September.

As the Wallace–Baruch affair continued in the press, the Soviets finally called for a vote on the Scientific and Technical Committee's report. The group was pleased by the Soviet vote in favour of the report, but the feeling was short-lived. The Soviet representative stated that his vote was accompanied by a reservation, based on the fact that the information on which the report's conclusions were based was incomplete and therefore should be regarded as hypothetical and conditional. Committee Two formally accepted the report of the Scientific and Technical Committee on 2 October and began hearing testimony from various experts in the field.

Although Committee Two was proceeding smoothly, various Soviet actions through October 1946 made it fairly clear that the sides were poles apart. Meanwhile, Baruch pressed Truman for an answer to his September letter that called for an early vote. By the time Baruch received permission in November to force a vote by the end of the year, the Baruch Plan had been all but rejected and his reputation subject to vicious attacks by the Soviets in the UN.

The Cold War steps in

On 13 November, at the first plenary meeting of the UNAEC in four months, the vote was ten in favour, two (USSR and Poland) abstaining, that the UNAEC should report its finding and recommendation to the Security Council by 31 December 1946. Despite Soviet delaying tactics, Baruch moved closer to his goal of an early vote. On 5 December, Baruch, whose position had been reaffirmed by the White House, proposed that the plan bearing his name be adopted as a recommendation to the Security Council, but did not insist on a vote on that day. On 20 December, the UNAEC rejected the Soviet proposal to postpone the vote for a week, while the Polish delegation proposed to refer the Baruch Plan to the Political and Social Committee of the UN General Assembly. At this point, Gromyko simply refused to participate any further, a position he maintained throughout the end of the year.

Several days later, on 26 December, Committee Two passed its report on safeguards and forwarded it to the Working Committee, which the next day discussed the Baruch Plan, one paragraph at a time. There was only one area of disagreement: the veto. The group agreed to report to the full UNAEC, with a cover letter explaining the remaining dispute, and a note that the Soviets had not participated. At the final meeting of the UNAEC, on 30 December, the group agreed to Baruch's proposal to adopt the Working Committee report and submit it to the Security Council the next day. It passed by a majority but without Soviet agreement, producing what future Democratic Senator from Connecticut, Joseph I. Lieberman, called 'a hollow victory' for the United States.

As planned, Baruch resigned shortly after the vote, giving his place to the US representative to the UN, Warren Austin, presumably strengthening the American hand by combining the negotiator and the representative in the same person. The Security Council discussed the report without much success until March 1947, when it passed a resolution to refer discussions back to the UNAEC. The UNAEC provided the second report in September; their deliberations had included twelve Soviet amendments to the first UNAEC report, all of which had been rejected. The Security Council did not consider the second report of the UNAEC, which continued to meet through the spring of 1948. A third UNAEC report concluded that the group had reached an impasse and requested that the Security Council suspend its deliberations. In the summer of 1948, the Soviets vetoed a Security Council resolution approving all the UNAEC reports, while a non-binding resolution of the General Assembly approved the majority plan, hoping that the UNAEC would one day find a way to bring atomic weapons under control. Hope apparently ran out in November 1949 when the General Assembly agreed to suspend the work of the UNAEC.

When Bernard Baruch presented the United States' initial proposal dealing with atomic weapons at the inauguration of the UNAEC in June 1946, he launched the first of what would become hundreds, if not thousands, of multilateral and bilateral discussions on arms-control measures during the next six decades. The Baruch Plan would have created an International Atomic Development Authority to control or own all activities associated with atomic energy, from raw materials to military applications, and inspect all other uses. The Soviets and other delegates challenged the US proposal since the Americans would not relinquish their atomic arsenal, while expecting others to forgo developing their own. They were not far off target. 'America can get what she wants if she insists on it', Baruch asserted in December 1946. 'After all, we've got it and they haven't, and won't for a long time to come.' Baruch was wrong on both points: the Soviets rejected his plan and soon produced their own atomic bombs (see Chapter 4).

'Neither the United States nor the Soviet Union was prepared in 1945 or 1946 to take the risks that the other power required for agreement', historian Barton Bernstein concluded. 'In this sense, the stalemate on atomic energy was a symbol of the mutual distrust in Soviet–American relations.' Washington's continued insistence, beginning with the Baruch Plan, upon intrusive inspection systems to verify treaty compliance, which Moscow viewed as sanctioned espionage, figured prominently in stalemating future arms-control endeavours. That it could have been otherwise in 'a struggle for the very soul of mankind', to quote former President George H. W. Bush, in a different but related context many years later, should not be surprising.

Chapter 4
Race for the H-bomb

Standing on the steps of 10 Downing Street on the afternoon of 23 September 1949, British Prime Minister Clement Attlee read a brief statement: 'His Majesty's Government has evidence that within recent weeks an atomic explosion has occurred in the USSR.' Apart from a call for greater effort towards international control of atomic weapons, the statement offered no further explanation. The announcement did not say when and where the explosion had taken place or how it had been detected, although it later came to light that the announcement came nearly a full month after the actual explosion—the test, of a plutonium type, had been conducted on 29 August—and had been detected after the fact by spy aircraft taking air samples.

None of that was revealed at the time, however. Journalists frantically trying to flesh out the story found other government officials equally tight-lipped. The public reception of the news was remarkably subdued. When the BBC led off its evening news broadcast with that announcement, the report was typically matter-of-fact. Across the Atlantic, President Harry Truman issued a similar statement more or less simultaneously. It, too, offered few details but tried to pre-empt a domestic political outcry with reassurances that the inevitability that the Soviets would someday develop the bomb 'has always been taken into account by us'. The implications were uncertain, but the message was clear. The American atomic monopoly was over sooner than most serious observers expected. For the British people, it was a reminder that their small, densely populated islands were highly vulnerable to the new weapons. For the American people, protected by time and space, the sense of imminent peril was always going to be less immediate.

Public surprise that the Soviets had perfected the bomb was remarkably muted. The development had come before nearly everyone expected it, but the capability was not in and of itself a cause of shock. Western forecasts for when the Soviets would cross the atomic threshold had varied widely, reflecting the dearth of hard evidence on the Soviet atomic programme. The first CIA estimate on the issue, dated 31 October 1946, predicted that the Soviets would produce a bomb 'at some time between 1950 and 1953'. Later estimates put greater emphasis on the latter end of that time span. Just five days before the Soviets exploded their first bomb, the CIA predicted that the 'earliest possible date' that the Soviets would be able to develop the bomb was mid-1950, but the 'most probable date' was mid-1953. Several policymakers contributed their own guesses. The American ambassador in Moscow, Walter Bedell Smith, who later became director of the CIA, told James Forrestal at the height of the Berlin blockade in September 1948 that it would be at least five years before the Soviets developed the bomb. 'They may well have the "notebook" know-how,' he told Forrestal, 'but not the industrial complex to translate that abstract knowledge into concrete weapons.' Sir Henry Tizard, head of the British Atomic Energy Programme, placed the date at 1957 or 1958. Some argued that it would be later. Others argued that the Soviets would never surmount the technological difficulties of the process. Even the worst-case scenarios envisaged by groups within the US air force projected that it would be 1952 or 1953.

The announcement was covered extensively in the world press, but generally the popular reaction was relatively calm. Some even used the absence of detailed information to question whether the Soviet explosion had really taken place. The public announcements of the bomb had refused to give any information on how the blast had been detected, which in turn fuelled claims from radical isolationists in Congress, such as Senator Owen Brewster (R-Maine), that the Soviet Union did not, in fact, have an atomic bomb. Doubters were aided by the absence of a follow-up performance by the Soviets. Not until two years later did the Soviets test their second atomic device. On 24 September 1951, the Air Force Atomic Energy Detection System picked up unusually intense acoustic signals within the Soviet Union, which were later confirmed to be another atomic explosion.

Re-evaluations of Soviet atomic capabilities in light of the news were that the Soviet stockpile would rise from about two a month to a total of about five or more a month by the end of 1950. That would reap, according to US intelligence estimates, a growth from a stockpile of approximately ten to twenty atomic bombs that the Soviets were likely to have by mid-1950, to about 200 by mid-1954. That figure constituted something of a critical threshold in American military planning. American defence planners had decided that once the Soviets had the capability to deliver approximately 200 atomic bombs to targets in the United States, they would be able to take out many of the most critical American targets and thereby inflict devastating blows to the US's war-fighting ability.

America's atomic monopoly

The United States moved surprisingly slowly in these early days to articulate a coherent strategic policy linking military planning to foreign policy objectives. For just over four years the United States had enjoyed an atomic monopoly. During that time, Washington, along with their closest transatlantic allies, especially the United Kingdom, had failed to craft a coherent doctrine that brought the awesome power of atomic weapons into the service of Western foreign policy, even as the consensus grew that the West was in a new kind of war with the Communist regime in the Soviet Union. All they could muster were relatively hollow threats on an ad hoc basis. It was an approach US Secretary of Defense James Forrestal complained was 'a patchwork job'. Having formally adopted the concept of the 'containment' of Soviet Communism in late November 1948, most policymakers within the Truman administration simply assumed, or perhaps hoped is a better word, that the American atomic monopoly would somehow intimidate the Soviets from breaches of the peace for fear of precipitating an all-out war.

But if that was Truman's intention, it did not appear to work. The bomb was supposed to be the 'winning weapon', but by 1948 it was abundantly clear that the West was neither winning the Cold War nor preventing Moscow from repeatedly challenging Western interests. The Soviets seemed to have the initiative on all the fronts that mattered. French strategist Raymond Aron wrote in 1954 that, 'When one surveys the entire period since the Hiroshima explosion, it is difficult to resist the impression that the United States has lost rather than gained by its famous atomic monopoly. It has been of no use in the Cold War.' The political crises just seemed to keep coming: Yugoslavia, Iran, Greece, Italy, France, and Germany. And underpinning the entire debate on whether the bomb would be used was the issue of whether or not it *could* be used.

Attention now turned to building a real atomic capability. Political pressure to bring the boys back home and create a smooth economic transition from war footing to peace led to a massive demobilization in the wake of World War II. There were higher domestic priorities than gearing up for another war. In the perennial guns-or-butter debate, guns lost out. For those most concerned about the emerging threat of the Soviet Union, such as James Forrestal, the demobilization went too far. Anxiety was becoming prevalent among American military planners that the post-war demobilization had left the United States military barely able to maintain its existing commitments; if the Soviets forced military action in another theatre, there were simply not enough Western forces to stop them. The limits were political, not economic or logistical. In contrast with the other great powers, the United States had emerged from World War II on a solid economic footing; its territory was unharmed and the fabric of its society was intact. All it took to reverse the weakening of American defence forces, critics of the Truman administration's low defence spending limits argued, was the political will to do so.

A by-product of the post-war demobilization was that the US atomic programme had nearly ground to a halt. In his announcement of the bombing of Hiroshima, Truman implied that atomic bombs were rolling off the assembly line: 'In their present form these bombs are now in production and even more powerful forms are in development.' While not technically incorrect, it was deliberately misleading. In fact, the Americans had only a handful of bombs then and through the early Cold War, the result of political decisions taken in Washington rather than any logistical limits. By the end of 1945 the United States had built only six atomic bombs; by 1947, only thirty-two; by 1948, 110. By the end of 1949, when the Soviets detonated their first atomic bomb, the United States had 235 weapons. The stockpile grew at much faster rates after 1950, when Truman authorized a massive military build-up on the back of the Korean War.

Building more bombs would accomplish little without devising a viable nuclear doctrine and declaratory policy. The first Cold War crisis to test these elements was the crisis in Berlin in 1948, the first of the Cold War's genuine nuclear crises. One observer claims that in view of the precedent it was setting, 'it is clear beyond any shadow of doubt that this was the most critical crisis of the Cold War'. When Stalin blockaded Berlin in mid-1948, it seemed to provide the tangible proof that was hitherto lacking from the warnings of Forrestal and others that not only did the Soviets have a conflict of interests with the United States, but they were also willing to act upon those interests. In response, Truman made a remarkable commitment to maintain the presence in Berlin, although he had little idea how he would accomplish this. The most famous response to the challenge was the Berlin airlift, an ingenious effort to supply the two million residents of the Western sectors of the city by air. But Truman never regarded the airlift as anything more than a delaying tactic.

The Joint Chiefs of Staff had made it abundantly clear that there was no way to win a conventional war in Europe against the Red Army. Although some top-secret American war plans tried to incorporate the use of atomic bombs, it remained unclear how the new weapons could best contribute to the effort. Military planners hoped that the atomic bomb would be a 'distinct advantage' in war with the Soviet Union, at the same time as recognizing that the geography and structure of the Soviet Union offered relatively few high-value targets. Targeting cities such as Moscow and Leningrad was logistically viable but offered many disadvantages with little gain—against a country that had lost in the order of 27 million lives in World War II, the shock value was likely to be muted and the move was unlikely to contribute to victory. World War II had shown the value in attacking the enemy's war-making potential with strategic air power, but the Soviet Union was very different from Japan or Germany. The Soviet transportation system, identified by planners as 'the most vital cog in the war machine of the USSR', spanned vast distances with relatively few dense hubs; it was simply too spread out to be a viable target for the relatively few atomic bombs the United States possessed during the period. Soviet military industries were also dispersed, and only the country's petroleum supplies appeared vulnerable to strategic bombing. Not until 1956 did the National Security Council believe that the United States had the capability to carry out a 'decisive strike' against the Soviet Union.

The post-war demobilization had seriously depleted the practicable options available to the president to exploit the atomic monopoly, and it was further hampered by the extreme secrecy surrounding information related to atomic weapons. Not even the president was able to get a straight answer on how many weapons were in the US stockpile and what they could do. Catalysed by the apparent military impotence revealed by the absence of any good options to deal with the Berlin blockade, the Joint Chiefs of Staff undertook to review the defence posture of the United States, beginning with nuclear strategy. Forrestal and the Joint Chiefs of Staff used the blockade in their efforts to thwart Truman's tight defence budgets, seizing the opportunity to argue that relying on a perception of strength was not enough; it had to be backed up by tangible military capabilities. At the height of the Berlin blockade, frustrated by Truman's reluctance to commit to 'whether or not we are to use the A-bomb in war', Forrestal took it upon himself to authorize the Joint Chiefs to base their war planning on the assumption that nuclear weapons *would* be used.

Furthermore, the blockade demonstrated the inadequacy of American nuclear strategy when Washington was forced to improvise an atomic deterrent by sending B-29 'atomic' bombers to Britain and Germany. It was a bluff.

Few had thought seriously about how to wage atomic war. Winston Churchill suggested presenting the Soviets with an ultimatum threatening that if they did not retire from Berlin, abandon East Germany, and retreat to the Polish border, US atomic bombers would raze Soviet cities. The US Commander in Germany, General Lucius D. Clay, took a similar line by telling Forrestal that he 'would not hesitate to use the atomic bomb and would hit Moscow and Leningrad first'. British Foreign Minister Ernest Bevin was also enthusiastic for the opportunity to show Moscow 'we mean business'.

As tempting as it was to lash out at Moscow, Washington was inclined to tread lightly. As official British government policy put it, it seemed doubtful that the West could add the 'scorpion's sting' to such nuclear threats, a point that US policymakers quietly conceded. That Stalin had provocatively blockaded Berlin in the first place, despite the American atomic monopoly, was clear evidence that a deterrent had to be manufactured and explicit; the mere existence of atomic weapons was not enough. Moreover, the United States, many feared, had made commitments that exceeded its military capabilities.

The Soviet bomb

Stalin publicly professed indifference to the deterrent effect of the bomb. It was a premonition of the wide gap between Soviet and American understanding of nuclear deterrence that became entrenched in following decades. 'The atomic bomb', he claimed in remarks published in *Pravda* in September 1946, 'is intended to frighten people with weak nerves, but it cannot decide the fate of a war'. Instead, he maintained an unshakeable faith that so-called permanent operating factors would ensure that the Soviet Union prevailed in any future war as they had in the last.

Stalin's calculated indifference was a strategic gambit. It was useful politically and diplomatically, but intentionally masked reality. Behind this public façade, Stalin's private comments and directions showed a more nuanced understanding of the potential impact of the atomic bomb on international relations. His own scientists had alerted him by May 1942 that the British and Americans might be jointly seeking an atomic bomb—in fact, he knew about the Manhattan Project even before Harry Truman did—but he was slow to grasp the import of the new weapon. He had been sceptical at first that such a weapon was significant; when his intelligence directorate informed him that some reports indicated that the British and Americans were collaborating on an atomic bomb, he voiced suspicions that it was part of a deliberate misinformation programme. Once convinced—paradoxically, by the suspicious absence of scientific information appearing in journals from Anglo-American government efforts to keep information from the

Germans rather than any positive confirmation—Stalin clearly grasped the significance of the bomb. Pavel Sudoplatov, a former Soviet spy, claims that when in October 1942 a senior Soviet scientist suggested simply asking Churchill and Roosevelt about the programme, Stalin responded that 'You are politically naïve if you think that they would share information about the weapons that will dominate the world in the future', a comment as interesting for its evident suspicion of his allies as for his recognition of the revolutionary potential of the atomic bomb.

The Soviets had started a bomb programme in 1943 through fear that the Germans might get to the bomb first, but the resources devoted to it fluctuated at a time when there were so many other pressing issues. It was, after all, a massive and expensive risk—only the United States had the luxury of territorial security, natural resources, and two billion dollars to spend on the programme. Only after Hiroshima did atomic weapons become a top priority.

Prior to then, Stalin seems to have grossly underestimated the scale of destruction wrought by the new weapon, though that doubtless changed with the dramatic evidence of the atomic bombings of Japan. But if there were any doubts that Stalin came to appreciate the potential of the bomb to alter international politics, these were dispelled with his orders to Soviet security chief Lavrenti Beria and the Soviet Union's leading atomic scientist Igor Kurchatov to spare no resources in ramping up the Soviet bomb programme 'on a Russian scale'. Stalin promised that the atomic scientists would be given unprecedented freedom in their work and all the material support the state could muster. 'Hiroshima has shaken the whole world. The balance has been broken', he told his scientists. 'Build the Bomb—it will remove the great danger from us.' It was a decision that had far-reaching effects on the development of a modern Soviet military-industrial complex, effectively laying the groundwork for his successors to build a massive nuclear programme that would establish practical strategic parity with the West within two decades.

Soviet spies played an important role. While the Manhattan Project devoted most of its early security resources to protecting against German espionage, the Soviets benefited from a steady stream of detailed information—including specific blueprints—spirited out of the programme by fellow travellers and agents such as Klaus Fuchs, David Greenglass, and Julius and Ethel Rosenberg (the last two were executed for treason in 1953) (Figure 6).

The opening of Soviet archives in the early 1990s, together with the declassification of the so-called VENONA transcripts—translations of some 3,000 messages sent between Moscow and Soviet intelligence stations in the US in the 1940s—paint a picture of a golden age of Soviet espionage. This information, in turn, was channelled directly to the Soviet scientists by Beria's organization. At the time, the beginning of the Cold War, few in the West doubted that this intelligence directly accelerated the Soviet bomb programme.

During the Stalin years, Soviet military doctrine basically ignored nuclear weapons as offensive weapons. But there were active efforts to defend against American long-range bombers that might be armed with atomic bombs. Around 1948, anti-aircraft defences were assigned a higher priority, around the same time that Soviet scientists and the Ministry of Defence first began looking into the technology of both intercontinental ballistic missiles and anti-ballistic missiles.

Stalin's views of the atomic bomb gradually changed. Combined with the tight secrecy imposed by the Soviet regime, efforts to determine whether the Soviet leader was deterred by the American bomb are complicated. A leading scholar of Soviet foreign policy, Vladislav Zubok, has argued that Stalin's thinking on nuclear matters, like that of most leaders in the nuclear club, evolved over time. Zubok speculated that:

6. Ethel and Julius Rosenberg leaving New York City Federal Court after arraignment. The couple were later convicted of espionage and executed.

If somebody had asked Stalin after Hiroshima in 1945 and again at the end of his life in late 1952, whether he believed the bomb would affect the likelihood of war in the future, he might have given two different answers. In 1945, he would probably have said that the US atomic monopoly encouraged America's drive for world hegemony

> and made the prospects of war more likely. In early 1950, after the first Soviet test, he was ready to say that the correlation of forces shifted again in favor of the forces of socialism and peace.

Winston Churchill insisted that America's atomic bomb was all that held Communist advance at bay. 'Nothing stands between Europe today and complete subjugation to Communist tyranny but the atom bomb the Americans possess', he told an audience in Wales in 1948. It was a refrain he repeated often.

'Years of opportunity', or not

In retrospect, it is surprising that the world's sole atomic power, the United States, did not make more aggressive moves to prevent others from developing the bomb. That is not to say that the idea of preventive war was not debated. It had long been discussed in classified circles. Some argued that the United States had squandered its advantage, that America's greatest military asset had been wasted, a decision that could have catastrophic consequences. James Forrestal wrote in late 1947 that the remaining years of the monopoly, however long that would be, would be the West's 'years of opportunity'. As early as January 1946, General Leslie Groves, the military commander of the Manhattan Project, reflected: 'If we were ruthlessly realistic, we would not permit any foreign power with which we are not firmly allied . . . to make or possess atomic weapons. If such a country started to make atomic weapons we would destroy its capacity to make them before it had progressed far enough to threaten us.' Nonetheless, the US government never came close to implementing a preventive war strategy and the most powerful government officials did not support the idea.

The sense of foreboding ran deep in policymaking circles about what the Soviets might do if they had the bomb, leading to a full range of prescriptions. Talk of preventive war was controversial, but held a mantle of respectability that peaked in the late 1940s and early 1950s, a respectability that faded rapidly in the midst of the thermonuclear revolution of hydrogen warheads and long-range ballistic missiles.

Although the American public remained decidedly cool to the idea of preventive war—various polls in the early 1950s pegged public support for preventive war against the Soviet Union at between 10 and 15 per cent—support for the idea of waging war on the Soviets before Stalin built up his own large atomic arsenal enjoyed remarkably wide, if publicity-shy, support in official Washington, and Moscow knew it.

The air force and the RAND Corporation acted as loci for the idea of preventive war and remained havens for it long after it had been discredited in other circles. But in the late 1940s and early 1950s, when there was still an apparent window of opportunity, support for preventive war also came from less expected quarters. Leading atomic scientist Leó Szilárd reportedly advocated preventive war as early as October 1945. George Kennan and fellow State Department Kremlinologist Charles Bohlen, both relative moderates in terms of Cold War military policy, found the logic compelling.

There were a number of reasons why such arguments never won the day. To begin with, it was a question of national character. America was not in the habit of starting wars. Having been on the receiving end of the surprise attack at Pearl Harbor, US policymakers—and the American public—presumably held US foreign policy to a higher standard. Although the United States had long reserved the right to take preventive military action, actually doing so would first have to overcome deeply held national convictions that starting wars was not the best way to behave in the international arena.

More important, though, were doubts that preventive war against the Soviet Union would be successful. The post-war demobilization placed severe limits on US military capabilities and the Western European allies were in no position to make any meaningful military contribution to the effort. The Red Army, which Stalin maintained in large numbers as his own form of 'deterrent', would have had a clear run to the English Channel. This in turn raised two questions: To be effective, would not preventive war have required more than air strikes with A-bombs? And would not the United States have been required to send in ground troops to occupy the Russian heartland? The plain fact was that the United States was neither capable nor inclined to wage a preventive war against the Soviet Union to prevent a Communist bomb.

The thermonuclear decision

Clearly, Moscow had not been awed by the American atomic monopoly. And now that that monopoly had been broken, many observers were convinced that the Soviets would become even more dangerous. Informed opinion, including the intelligence community, recognized that it would still take time for the Soviets to develop a usable stockpile—by 1950 the Soviets had approximately five atomic weapons to the United States' 369. The United States faced two potential paths. One was to seize the opportunity to push for bilateral disarmament. The Soviets had baulked at early efforts at international control of atomic weapons on the basis that they would be relinquishing the right to develop their own atomic capability while the United States retained its arsenal. Now that both powers had the bomb, it would in effect be a mutual sacrifice. The other potential path was to engage in full-scale competition and an arms race. For a variety of reasons, mostly derived from the Cold War mindset, the administration chose the latter course. It was a watershed moment.

Nevertheless, hawks inside government continued to push their agenda. James Forrestal had long complained that the tight budget ceilings imposed by President Truman were forcing 'a minimum, not an adequate strategy'. His successor,

Louis Johnson, was ideologically inclined towards fiscal restraint and not overly inclined towards challenging his commander in chief's budget directives. Given the string of Cold War setbacks—especially the Soviet atomic test, and the 'loss' of China to Mao Zedong's Communist Party, both in 1949—political pressure eventually pushed Truman towards reconsidering defence spending and the strategy to go along with it. By the end of the process, defence spending increased by 458 per cent by the fiscal year 1952 over the budget for the fiscal year 1951, and the level of manpower in the Defense Department was raised to nearly five million from a 1951 level of 2.2 million.

During the winter of 1949–50, a highly classified debate had been raging in defence and scientific circles over whether to proceed with a new generation of weapon, this one exploiting the energy released when hydrogen atoms were fused rather than split, as they were in an atomic bomb. The new kind of weapon—variously termed a hydrogen, thermonuclear, or just nuclear bomb—was informally dubbed 'the super', a reference to its potential to dwarf even the explosive power of an atomic bomb. Preliminary research into such a weapon had been undertaken within the Manhattan Project by a team of scientists led by physicist Edward Teller. But with no hope of immediate success and with military budgets shrinking in the post-war economic environment, the research was halted. Based on theoretical data, Teller predicted that a hydrogen bomb would be several hundred times more powerful than the Hiroshima bomb, capable of devastating an area of hundreds of square miles, with radiation travelling much farther.

The debate centred on whether such a weapon was needed, the morality of its manufacture, and the impact its development would have on relations with Moscow. Producing a bitter mood, it eventually split not only the policymakers but also the atomic scientists themselves. In January 1950, Truman received a delegation headed by Dean Acheson, now secretary of state, which advocated development of the hydrogen bomb. After a meeting lasting only seven minutes, the president decided to press ahead with the research, despite the fact that there was no hard evidence that the hydrogen bomb would ever become a reality, and a number of scientists claimed that it couldn't be done. Many more, including James Conant and J. Robert Oppenheimer, the physicist who had led the Los Alamos team during the Manhattan Project, argued that it was unnecessary. Even Albert Einstein came out publicly against developing the hydrogen bomb: 'The idea of achieving security through national armaments is, at the present state of military technique, a disastrous illusion . . . The armament race between the USA and the USSR, originally supposed to be a preventive measure, assumes hysterical character.'

The Atomic Energy Commission's own advisory committee emphasized that the hydrogen bomb lent itself to genocide but not much else:

> The use of this weapon would bring about the destruction of innumerable human lives; it is not a weapon which can be used exclusively for the destruction of material installations of military or semi-military purposes. Its use therefore carries much further than the atomic bomb itself the policy of exterminating civilian populations.

Truman's statement announcing his directive betrayed none of the drama of the top-secret debate behind the scenes. In a brief, spare statement that included the usual call for greater international control of atomic arms, Truman announced that:

> It is part of my responsibility as Commander in Chief of the Armed Forces to see to it that our country is able to defend itself against any possible aggressor. Accordingly, I have directed the Atomic Energy Commission to continue its work on all forms of atomic weapons, including the so-called hydrogen or superbomb. Like all other work in the field of atomic weapons, it is being and will be carried forward on a basis consistent with the overall objectives of our program for peace and security.

It was a momentous decision, paving the way for the thermonuclear revolution and the arms race that went along with it.

The sense of urgency forced quick action. A few weeks after Truman's announcement, Louis Johnson, at the prompting of the Joint Chiefs of Staff, requested 'immediate implementation of all-out development of hydrogen bombs and means for their production and delivery'. By early March 1950, the thermonuclear weapon programme had been ramped up to 'a matter of the highest urgency'.

The same day that Truman authorized development of the hydrogen bomb, he instructed Acheson and Louis Johnson to reassess the Soviet threat in light of the Soviet Union's nascent atomic capability and recent Cold War developments. Under the direction of Paul H. Nitze, Kennan's successor as director of the State Department's Policy Planning Staff, a group of state and defence officials formulated a comprehensive statement of a national security strategy and submitted it to the president in early April 1950. Known by its bureaucratic designation as NSC 68 'United States Objectives and Programs for National Security', the document was deliberately alarmist and made the case for a massive build-up in resources and a hardening of strategy to go along with it. With its urgent tone and blunt, hawkish policy prescriptions, the document reflected a change in direction in policy terms, but its substance expressed the mood of many Washington policymakers that had been brewing for some time.

NSC 68 was fundamentally concerned with the problem of 'weapons of mass destruction' (the first to introduce the term to policy documents). It estimated that 'within the next four years, the USSR will attain the capability of seriously damaging vital centres of the United States, provided it strikes a first blow and provided further that the blow is opposed by no more effective opposition than we now have programmed'. It warned that once the Soviet Union 'has a sufficient atomic capability to make a surprise attack on us, nullifying our atomic superiority and creating a military situation decisively in its favor, the Kremlin might be tempted to strike swiftly and with stealth'. In these circumstances, and estimating the prospects of the international control of atomic energy as negligible, Nitze and his associates suggested that the United States had little choice but to increase its atomic and, if possible, its thermonuclear capabilities as rapidly as it could. The atomic stockpile should be rapidly increased and the hydrogen bomb programme continued at a greatly accelerated pace.

NSC 68 also warned of the dangers of 'piecemeal aggression' whereby the Soviets could threaten American interests without resorting to direct military confrontation. By exploiting Washington's unwillingness to use its atomic weapons unless directly attacked, Moscow might pose a military threat by other, more abstruse methods, which could potentially throw American defence policy into disarray and bypass whatever limited effect the atomic deterrent might be having. When North Korean troops marched on South Korea on 25 June 1950, at the height of the internal administration debate over NSC 68, it posed what was in many ways a novel challenge; it was not a scenario anticipated by existing Western strategy. In the words of Raymond Aron, 'The Korean War had taught world leaders that there are more things in heaven and earth than in models.' The Soviet preponderance of conventional military forces, compounded by an incipient atomic capability, which included a 'probable fission capability and possible thermonuclear capability', posed a serious challenge for which military planners strove to account. Consequently, it provoked a comprehensive reappraisal of US national security assumptions and seemed to lend weight to arguments for embracing NSC 68.

Beyond the realm of logic

The decision had at once profound effects on nuclear weapons development and nuclear policy. The atomic arsenal received new emphasis, with American science and technology engaged in producing smaller and cheaper atomic warheads that permitted the US army to deploy thousands of tactical atomic weapons on the battlefield. Nuclear research and development was boosted by the desire of each branch of the armed services for a piece of the action. During the 1950s, the army turned its attention to intermediate-range, land-based ballistic missiles, and the navy, first to aircraft-carrier-based atomic bombers and then to nuclear powered and armed submarines. But the mainstays of US strategic forces continued to be the bombers of Strategic Air Command. More importantly, work was accelerated on the H-bomb project, and on 31 October 1952 the United States detonated its first thermonuclear device, in the Pacific.

The explosion was the culmination of an extraordinary effort on the part of the Truman administration to maintain its ascendancy over the Soviet Union on the nuclear ladder and provided a watershed for deterrence. With the opening phases of the thermonuclear revolution now a reality, policymakers struggled to comprehend the scale of destruction of the new technology. Edward Teller had predicted in 1947 that the new weapon would be capable of devastating an area of 300 or 400 square miles and that radiation could well travel much farther. In terms of military strategy, such a regional scale clearly changed the whole nature of the weapon. But it didn't take long to grasp that such a weapon might well transform the nature of war and peace themselves. As Churchill put it, 'The atomic bomb, with all its terror, did not carry us outside the scope of human control or manageable events in thought or action, in peace or war. But . . . [with] the hydrogen bomb, the entire foundation of human affairs was revolutionized.'

While recognition of this exacerbated psychological gap between strategic weapons and victory prompted a sharpened focus of strategic thought that lasted at least a decade and a half, US policymakers were forced to deal with its consequences on a more immediate level. Seasoned war leader Eisenhower declared that with the existence of employable thermonuclear weapons, 'War no longer has any logic whatever.' And to prove the point, the Soviet Union successfully detonated its first thermonuclear device less than a year later, on 12 August 1953; it was a limited explosion about twenty-five times smaller than the US effort. In November 1955, the Soviet successfully air-dropped an H-bomb, with an explosive power of 1.6 megatons.

Great Britain joined the atomic club on 3 October 1952, with a successful test near the Monte Bello Islands, off the coast of Australia, and the thermonuclear club on 15 May 1957, with an H-bomb explosion of 200–300 kilotons, at Christmas Island in the Pacific. Under the relentless guidance of Charles de Gaulle, France acquired its own nuclear strike force with a test in the Sahara Desert in Algeria in 1960, followed by a thermonuclear explosion at Fangataufa Atoll, South Pacific, in 1968. Fearful of both superpowers and with an eye on India, China joined the nuclear club in 1964 and thermonuclear club in 1967, with a bomb that was dropped over the Lop Nor test site.

During the late 1960s, Israel, under the initial direction of the 'father' of the French atom bomb, Francis Perrin, who built the Dimona Nuclear Research facility, became the sixth nation with nuclear weapons capability, although the Israeli government denies it. India (1974) and Pakistan (1998) became the seventh and eighth nations to acquire nuclear status, focusing attention on their great rivalry in South Asia. And North Korea joined the nuclear club in October 2006.

During the 1970s, South Africa's Atomic Energy Board established a nuclear weapons programme. Using largely open sources, they enriched uranium. In August 1977, a Soviet satellite discovered South Africa's nuclear test site in the Kalahari Desert; however, under pressure from the United States, the USSR, and France, South Africa temporarily postponed its plans until 1982, by which time it had developed its first complete nuclear device. Then, for reasons very much of its own, one suspects, South Africa closed down its nuclear weapons programme and dismantled its weapons facilities in 1989. Two years later, it joined the Nuclear Non-Proliferation Treaty.

While opposition to nuclear energy first emerged shortly after the atomic bomb was built, significant anti-nuclear opposition did not emerge until the 1950s. The American hydrogen bomb test on Bikini Atoll in March 1954 made the world acutely conscious of radioactive fallout for the first time. Fallout from the explosion rained down on the Marshall Islanders and a Japanese fishing boat, the hapless *Lucky Dragon*. Shortly afterwards, a handful of London housewives started a campaign to pressure the US government to stop its nuclear testing, and this became the beginning of the test ban movement which provided the drumbeat and groundwork for the Comprehensive Nuclear Test Ban Treaty four decades later. The initial protest later became the National Campaign for Nuclear Disarmament, of which British philosopher-mathematician Bertrand Russell was the guiding spirit. If war no longer had any logic, nor did the further testing of nuclear weapons.

Chapter 5
Nuclear deterrence and arms control

When, at the end of the 1970s, Queen Elizabeth II declared that nuclear weaponry's 'awesome destructive power has preserved the world from a major war for the past thirty-five years', she reflected an opinion held by most Cold War statesmen and, subsequently, by many academics. Later, historian John Lewis Gaddis viewed the forty-five-year Cold War as 'the long peace' since there were no direct major hostilities between the United States and the Soviet Union. It was an unprecedented accomplishment, he argued, as 'prior to that moment, improvements in weaponry had, with very few exceptions, increased the costs of fighting wars without reducing the propensity to do so'. In this sense, then, the nuclear revolution was akin to a great earthquake, setting off a series of shockwaves that gradually worked their way through the political system.

But not all observers agreed. Some suggested that nuclear weapons were 'essentially irrelevant' to keeping the peace because, even without these new destructive devices, a world war had become too costly for a rational leadership to engage in it. A former State Department official, Raymond L. Garthoff, acknowledged that the existence of nuclear weaponry in the hands of both superpowers undoubtedly exercised 'a restraining, deterring, effect'. But had nuclear weapons not existed, he concluded, 'it remains highly probable that neither the United

States nor the Soviet Union would have attacked the other, and less certain but also probable that neither would have taken other military actions so provocative as to have precipitated general war between the two powers'.

There is little likelihood of agreement on the general proposition that the destructive power of nuclear weapons maintained a relative peace between the superpowers. But an important caveat should be inserted. In 1985, for example, Lord Carrington, the Secretary General of the North Atlantic Treaty Organization (NATO), stated his belief in the value of deterrence: 'I don't believe it's worked; I *know* it's worked. There hasn't been a war for 40 years . . . there is [no] other way at the present time of keeping the peace for the world.' In referring to 'keeping the peace for the world', he was speaking of the absence of a nuclear war, since non-nuclear powers continued to wage conventional war freely, though nuclear powers less freely.

Wars fought with conventional weaponry were a common occurrence during the Cold War and could be waged by non-nuclear powers with little restraint. Nuclear powers could wage a limited conventional war, but they were restrained from fighting one another. Case studies of Cold War-era conflicts suggest two ironclad, unwritten rules: first, no nuclear power may use military force against another nuclear power; and, second, a nuclear power, using military force against a non-nuclear nation, may not use nuclear weapons. Moreover, possessing nuclear weapons did not necessarily deter a non-nuclear nation from waging war with a client state of a nuclear power, as the United States found out in the Korean and Vietnam Wars.

Evolution of nuclear deterrence

Not until the second decade of the nuclear age was the danger of nuclear weaponry and the perception of this danger sufficient to give rise to the concept of deterrence and create a Cold War stalemate. Eugene Rabinowitch, editor of the *Bulletin of the Atomic Scientists*, chose 1956 as the birth date of the 'Age of Deterrence', calling it AD 1, 'the first year of deterrence'. Subsequently, others dated its arrival from 1954, 1955, or 1957. The *Random House Dictionary* (1987) chose 1955 as the date of its appearance, and defined it as 'The distribution of nuclear weapons among nations such that no nation will initiate an attack for fear of retaliation.' The standoff was also known as the 'balance of terror', a phrase made famous by Winston Churchill, but this was a bit too stark for popular consumption, while the term 'deterrence' was more easily digested (see Box 1).

With the advent of thermonuclear devices (H-bombs) and the introduction of nuclear-tipped, long-range ballistic missiles by the late 1950s, the concept of nuclear deterrence gained widespread currency. As the nuclear arsenals expanded in the 1960s, the phrases 'deterrence policy' and 'deterrence strategy' were used as euphemisms for 'nuclear policy' (short for 'nuclear weapons policy') and 'nuclear strategy'. And strategic theorists gradually linked words such as 'credible', 'effective', 'stable', and 'mutual' to the concept of a nuclear balance or deterrence.

These theorists also speculated about possible methods of employing the expanding nuclear arsenals. A 'first strike' could take place when a nation thought it had sufficient nuclear forces to overwhelm its foe and thus achieve victory, while a closely related 'pre-emptive strike' would call for launching a nuclear strike when a nation anticipated its enemy was preparing a first strike. A 'retaliatory strike' or 'second-strike' capability referred to a nation's ability to absorb a nuclear first strike and still retain sufficient weapons to inflict unacceptable damage on its attacker, or at least what it hoped would be unacceptable.

Policymakers and the public, however, rarely saw strategies in such stark forms. Thus deterrence emerged as neither a military strategy nor policy; it was simply recognized as a political reality.

Box 1 Stages of weapons development

Research and development (R&D): this period can take from a year or two to more than ten years, during which concepts and basic technologies are explored.

Engineering and manufacturing development (EMD): it can take five years or more to engineer and develop the industrial processes to manufacture and assemble a system.

Developmental testing: this is conducted throughout the R&D and EMD phases to learn about the strengths and weaknesses of the new system and to apply these technologies in a military environment.

Operational testing: this is conducted with production equipment in realistic operational environments—at night, in bad weather, against realistic countermeasures.

Production: initial quantities are usually small and later, after successful operational testing, a system may go into 'full-rate production'.

Deployment: the fielding of the new system, either in large or small quantities, in military units to develop or enhance tactics, techniques, and procedures for the use of the system if that has not already been done in the development phase.

When the US and USSR governments believed their military services were able to absorb a nuclear first strike and still possess sufficient forces for retaliatory strikes—as they did by the end of the 1960s—mutual deterrence had arrived, in fact if not in formal policy.

If deterrence gradually became mutual, the perceptions and policies of the two superpowers had diverged at the very onset of the Cold War. Their socio-political systems, grounded on differing ideological, geopolitical, economic, and political ambitions, created serious concerns about the designs and intentions of each other. 'For more than four decades,' Strobe Talbott lamented in *Time* magazine, 'Western policy has been based on a grotesque exaggeration of what the USSR could do if it wanted, therefore what it might do, therefore what the West must be prepared to do in response.' This led to grossly exaggerated worst-case assumptions about Soviet capabilities. At the same time, a disturbing change had begun to take place in the United States as militarism insinuated itself into American life. The scepticism about arms and armies that informed American society from its founding had started to vanish. Political leaders—liberals and conservatives alike—became enamoured of military might. The Soviet ambassador to Washington, Anatoly Dobrynin, acknowledged in his memoirs that Moscow's Cold War policies were also unreasonably dominated by ideology, and this produced continued confrontation. The superpowers, Mikhail Gorbachev later concluded, had been mesmerized by ideological myths.

These ideological and political tensions resulted in the adoption of different strategies to avoid a nuclear showdown. As a consequence, the United States addressed the problem of preventing war almost exclusively in terms of military capabilities. The Soviet Union, for its part, addressed preventing war primarily in terms of political motivations and intentions. The different focus of the two powers had important effects on the military doctrines and forces of each.

Throughout the Cold War, American leaders usually pursued a nuclear strategy that was, in the end, contradictory. For example, President Harry Truman was convinced that, on the one hand, nuclear weapons played an essential role in the democratic world's defence against its enemies, but, on the other hand, he feared that a war involving nuclear weapons most likely would destroy the US and modern civilization. In his January 1953 farewell address, Truman declared that 'starting an atomic war is totally unthinkable for rational men'. This was, he later stated, 'because it affects the civilian population and murders them by wholesale'. President Dwight Eisenhower would come to view war with thermonuclear weapons as 'preposterous'. Yet as these and subsequent administrations acknowledged that nuclear war was 'unthinkable', American political leaders and military chiefs continued to seek nuclear arsenals that might advance their more limited political objectives.

The Truman administration sought to tie the idea of deterrence to a way of enforcing the new policy of containment that was intended to prevent—and eventually reverse—the indirect and direct expansion of Soviet domination and influence. The administration's basic national strategy of containment sought not only to 'block further expansion of Soviet power', but 'by all means short of war' to 'induce a retraction of the Kremlin's control and influence . . . to check and to roll back the Kremlin's drive for world domination'. Washington hoped its atomic monopoly might expand the theory of deterrence (preventing a nuclear attack on the US) to include the possibility of 'compellence' (forcing a Soviet withdrawal from Eastern Europe).

The destruction of Hiroshima had little deterrent effect on Moscow, but it did prompt Stalin to insist upon Russia's possession of nuclear weaponry to maintain the balance of power. And, he looked differently on the Soviet expansion into Eastern Europe—Stalin saw it as creating a barrier against any future German ambitions as well as the restoration of Russia's historic borders.

During the early Cold War years there were a few US efforts to apply 'atomic compellence', that is, threaten nuclear war to achieve the desired outcome. Truman asserted in his memoirs that America's atomic monopoly had pressured Moscow's withdrawal from northern Azerbaijan in March 1946. Subsequent documents indicated that the Soviets were not moved by the threats.

In secret discussions during the crises of 1953–5, President Dwight Eisenhower insisted that the use of atomic weapons 'was neither unthinkable nor unwinnable'. When he implied a willingness to employ conventional and nuclear force to resolve issues arising from the Korean armistice, Indo-China, and the Nationalist Chinese offshore islands, Eisenhower was persuaded that Moscow would not intervene to aid China or escalate a local conflict that risked a confrontation with the US's superior nuclear forces. In an effort to get 'more bang for the buck', Eisenhower launched his 'New Look' programme that trimmed funds for the army and navy, while increasing monies allotted to expand the Strategic Air Command and increase America's nuclear arsenal.

Secretary of State John Foster Dulles's infamous 1954 essay in *Time* magazine, 'A Policy of Boldness', further embellished the administration's effort at 'atomic compellence'. Regional allies must be supported by 'massive retaliatory power', he argued. 'The way to deter aggression is for the free community to be willing and able to respond vigorously at places and with means of its own choosing.' It is uncertain that any of these threats altered Soviet or Chinese decision policies; but it certainly upset many in the foreign policy public who pointed out that the major Communist regimes had limited influence in many local disputes such as the Indo-China conflict.

Subsequently, three developments alarmed the American public and challenged Eisenhower's defence policies. On 22 November 1955, the Soviets surprised the administration by detonating an H-bomb; in August 1957, they tested an intercontinental ballistic missile; and in October, the Soviets startled the world by launching Sputnik I, the first orbiting artificial satellite. Public unease persuaded the president to create a commission, led by Rowan Gaither, to assess the nation's vulnerability. The Gaither report, titled 'Deterrence and Survival in the Nuclear Age', released on 7 November 1957, held that the Soviets would have a dozen operational intercontinental ballistic missiles within a year, while it would take the US two or three years to catch up—creating a 'missile gap' (President John F. Kennedy quickly learned it was the Soviets who faced a 'missile gap').

In July 1958, Eisenhower was presented with two alarming scenarios: in the first, a Soviet nuclear strike that 'wiped out' the federal government and destroyed the nation's economy; in the second, the Soviets destroyed all Strategic Air Command bases and still wreaked havoc on the nation. In the US's retaliation the Soviets would suffer approximately three times the US damage, but American losses were staggering at nearly 65 per cent of a population of nearly 178 million souls. Stunned, Eisenhower's views changed dramatically—in a general war, he concluded, there could be no winners—thus thermonuclear weaponry could only be used to deter.

Mutual assured destruction (MAD)

The policy of massive retaliation was formally replaced in September 1967 by Secretary of Defense Robert McNamara's recognition that the Soviet nuclear build-up was approaching parity, thus creating a situation of 'assured destruction' (critic Donald Brennen added 'mutual' to get the acronym of MAD). The idea of MAD did not sit well with American military chiefs preaching peace through strength. The 'first principle of deterrence', General Thomas S. Powers wrote in 1965, was 'to maintain a credible capability to achieve a military victory under any set of conditions or circumstances'. An angry air force general, Curtis LeMay, insisted that 'The deterrent philosophy we now pursue has drained away our red military blood.'

Nonetheless, with their budgets at stake, the US military devised a formula (the triad) that provided each service with a strategic function. The air force possessed strategic bombers and nuclear-tipped intercontinental ballistic missiles (ICBMs), the navy its submarine-launched ballistic missiles (SLBMs), and the army its intermediate-range ballistic missiles (IRBMs), nuclear artillery, and mines, as well as anti-missile defences. In theory at least, the nuclear triad reduced the chances that an enemy could destroy all of a country's nuclear forces in a first-strike attack, ensuring that a devastating second-strike response could be carried out (see Box 2).

Despising notions of parity and sufficiency, defence analysts and military chiefs sought to find a way to employ nuclear weapons and a reason to expand their arsenals. For brief exhilarating moments, they tossed about the ideas of nuclear war-fighting—limited nuclear war, 'graduated deterrence', 'essential equivalence', launch on warning, pre-emption, etc.—only to be dismissed. For example, the London *Economist* found 'graduated deterrence' had two fatal defects. First, 'the deterrent, because of being graduated to the scale of aggression, would lose some of its power to deter'. Second, if the 'deterrent' were used in a limited way, the self-restraint would not be recognized as such.

Since the Soviet military did not receive nuclear weapons until 1954 and did not have sufficient delivery systems for many more years, Moscow could not rely on nuclear deterrence. Thus, the Soviet approach to averting war was basically political. In contrast to the American focus on deterrence as the essence of strategy and policy, successive Soviet leaders reacted to the nuclear age by adjustments of strategy, policy, and even ideology to give the highest priority to preventing war.

In the immediate post-World War II years, Stalin did not see the Americans and British as embarking on military conquests, and he believed he could occasionally probe the West's determination without provoking a general war. But he did miscalculate in permitting the North Koreans to attack the South and in seeking to pressure the West to withdraw from Berlin. Nonetheless, during these pre-nuclear years Soviet military plans appear to have been mainly defensive.

Box 2 Ballistic missile basics

Ballistic missiles are classified by the maximum distance that they can travel, which is a function of the missile's engines and the weight of the missile's warhead. To add more distance to a missile's range, rockets are stacked on top of each other in a configuration referred to as staging.

There are four general classifications of ballistic missiles:

- *Short-range ballistic missiles*, travelling less than 1,000 kilometres (approximately 620 miles).
- *Medium-range ballistic missiles*, travelling between 1,000 and 3,000 kilometres (approximately 620–1,860 miles).
- *Intermediate-range ballistic missiles*, travelling between 3,000 and 5,500 kilometres (approximately 1,860–3,410 miles).
- *Intercontinental ballistic missiles* (ICBMs), travelling more than 5,500 kilometres.

Short- and medium-range ballistic missiles are referred to as theatre ballistic missiles, ICBMs are described as strategic ballistic missiles.

All ballistic missiles have three stages of flight, and short- and medium-range ballistic missiles may not exit the atmosphere or have a warhead that separates from its booster:

- The *boost phase* begins at launch and lasts until the rocket engines stop firing and pushing the missile away from earth. Depending on the missile, this stage lasts between three and five minutes. During much of this time, the missile is travelling relatively slowly, although towards the end of this stage an ICBM can reach speeds of more than 24,000 kilometres per hour. The missile stays in one piece during this stage.

Continued

Box 2 Continued

- The *midcourse phase* begins after the rockets finish firing and the missile is on a ballistic course towards its target. This is the longest stage of a missile's flight, lasting up to twenty minutes for ICBMs. During the early part of the midcourse stage, the missile is still ascending towards its apogee, while during the latter part it is descending towards earth. It is during this stage that the missile's warhead(s), as well as any decoys, separate from the delivery vehicle.
- The *terminal phase* begins when the missile's warhead re-enters the earth's atmosphere, and it continues until impact or detonation. This stage takes less than a minute for a strategic warhead, which can be travelling at speeds greater than 3,200 kilometres per hour.

Stalin's successors brought deterrence into their considerations: from the mid-1950s in theory, the mid-1960s in interim real capability, and the early or mid-1970s in terms of rough parity. After the Soviets exploded a hydrogen device, Prime Minister Georgy Malenkov was the first leader to warn that a nuclear war would mean the end of world civilization. Political opponents—such as Nikita Khrushchev—denounced him for repeating Eisenhower's warning, but as these critics succeeded him they soon sounded the same message.

In the late 1950s and early 1960s, Moscow did seek to brandish its nuclear weaponry, but it was paradoxically at the time of their greatest relative weakness. From the Suez Crisis in 1956 to the Cuban Missile Crisis in October 1962, Khrushchev attempted to turn the Soviet Union's weaknesses into a deterrent and even a political compellent by an outrageous, deceptive exaggeration of its nuclear capabilities. When Khrushchev decided to deploy Soviet medium-range ballistic missiles, IRBMs, tactical nuclear weapons, and nuclear-capable medium-range bombers secretly in Cuba, where they would be positioned to strike most of the continental United States within minutes, his reasoning was to bolster the Soviet deterrent. Whether he wanted to use this deterrent in an offensive or defensive role has been debated by scholars ever since. Once the deployments were discovered, John F. Kennedy responded to the challenge by implementing a naval blockade of the island and threatening military action if the missiles and bombers were not removed. After a week-long standoff, during which Strategic Air Command's forces were on airborne alert, the Soviet leader agreed to remove the missiles and a month later agreed to remove the bombers. After this spectacular failure, Moscow gave up attempting to achieve political gains from a marginal nuclear arsenal. For even as the Soviets built up real nuclear forces in the 1960s and 1970s and maintained parity in the 1980s, buttressing nuclear deterrence, they never again attempted to redress the nuclear balance by force—or even the threat to use force.

What little is known of Soviet war planning (and US planning) during the Cold War points to its armed forces seeking to prevail should deterrence fail. In 1955, Marshal Pavel Rotmistrov advanced a shift in Soviet nuclear doctrine to prevent a surprise attack from crippling their retaliatory forces by endorsing a pre-emptive strike (five years after the Truman administration advanced the same concept) when an imminent enemy nuclear attack was detected. In the bomber era, he stressed, the idea of pre-emption was *not* a cover for a surprise attack or preventive war: 'The duty of the Soviet armed forces is to not permit surprise attack by the enemy on our country, and in case an attempt is made, not only to repulse the attack successfully, but also to deal to the enemy simultaneous or even pre-emptive surprise blows of terrible crushing power.'

The pre-emptive doctrine was replaced in the late 1960s by launch under attack and, probably, in the 1980s by launch in retaliation.

In Washington, a debate persisted throughout the Cold War as to whether the Soviets were really prepared to accept the idea of deterrence or whether they were developing the weapons and strategy to go beyond 'defensive' deterrence. Soviet leaders, however, did not view the American concept of deterrence as either benign or defensive (as Washington did); instead they saw it as offensive—compellent and intimidating.

A question arises from this review of deterrence: How much is enough? British Labourite Denis Healey, shadow foreign secretary in the early 1980s, declared that only 5 per cent of the warheads in hand were actually necessary to deter Moscow, the remaining 95 per cent were merely to assure the general public. In its May 1992 issue, the *Bulletin of the Atomic Scientists* asked a group of specialists on nuclear topics: What is to be done with nuclear weapons? All wanted 'deep cuts' in the existing nuclear arsenals and most agreed nations should maintain 'the least amount [needed] to deter'. Many placed the desirable number of weapons to be retained at one hundred. Clearly, much that was done in the name of deterrence of a potential adversary was really done to provide reassurance to one's friends and allies and one's own people.

Arms control and nuclear stability

Arms races, according to conventional wisdom, were the result of conflicting foreign policy objectives and would, with a reduction in international political tensions, fade away. This historically grounded observation lost much relevance in the 1960s, when ICBMs with nuclear-tipped warheads turned the proposition on its head. Instead of military force supporting foreign policy, managing nuclear weaponry became a major foreign policy objective. Often shrugged off as arcane and obtuse discussions, these post-1945 arms-control negotiations played an important, but frequently overlooked, role. During the Cold War, arms control became the principal conduit for Soviet–American relations, and even in times of tension arms control trudged ever on in some form or another.

Arms control and disarmament policies were advocated during the Cold War for several purposes: to enhance the nation's security, reduce military expenditures, influence international public opinion, and gain a domestic partisan political advantage. However, the overriding reason the superpowers engaged in protracted negotiations that led to many agreements was the necessity in the nuclear era of maintaining a stable international environment.

Bernard Baruch's ill-fated effort to deal with atomic weapons at the inauguration of the United Nations Atomic Energy Committee in June 1946 (discussed in Chapter 3) launched the first of what would become hundreds, if not thousands, of multilateral and bilateral discussions on arms-control measures during the next four decades. Washington's continued insistence since then upon intrusive inspection systems to verify treaty compliance, which Moscow viewed as sanctioned espionage, figured prominently in stalemating future arms-control endeavours. Some critics have argued, with considerable justification, that verification issues have become excessively prominent in arms-control negotiations; they also argue that the US's demands were purposely designed to impede such negotiations or, if agreed to, would greatly enhance its opportunities for general intelligence gathering.

The arms-control activities shifted to more limited, technically oriented objectives in the 1950s, as radioactive fallout from atmospheric nuclear tests aroused worldwide efforts to halt the testing. President Eisenhower asked technical experts to develop a verification system, a move that had unexpected long-term results, since experts often complicate issues to a point where they become insoluble. After developing techniques that could distinguish between earthquakes and virtually all underground nuclear explosions, technicians kept searching to reduce the already quite low error rate. It became impossible to negotiate a comprehensive test ban because critics argued that one could not be *absolutely* certain that no cheating was going on. This overemphasis on technical details, in fact, made the problem of verifying the test ban seem more and more formidable, because the verification system demanded by the seismologically expert American politicians was always going to be too intrusive to be acceptable to the Soviet Union.

While Eisenhower obtained only an informal moratorium on testing, John F. Kennedy entered the presidency determined to negotiate a comprehensive test ban. When ambassador W. Averell Harriman went to Moscow to finalize the test ban in July 1963, a dividend of the resolution of the Cuban Missile Crisis, he took scientific advisers with him but deliberately excluded them from the negotiating team, emphasizing that arms-control negotiations were, fundamentally, political undertakings. As he later explained, 'The expert is to point out all the difficulties and dangers . . . but it is for the political leaders to decide whether the political, psychological and other advantages offset such risks as there may be.' By this time, however, it was all but certain that such a treaty was beyond reach. In addition to Senate soundings that a comprehensive treaty would not pass muster, Khrushchev had repeated his objections to its requirement of on-site inspections—the Soviet Union would never 'open its doors to NATO spies'.

Khrushchev indicated, however, that he would be willing to conclude a limited or partial test ban treaty. Accordingly, after ten days of tense negotiations, closely monitored and supervised by President Kennedy himself, the Treaty Banning Nuclear Weapons Tests in the Atmosphere, Outer Space, and Under Water—the Limited Test Ban Treaty—was initialled in Moscow by the principal negotiators on 25 July 1963.

When the Soviet Union achieved a crude parity in strategic weaponry in the late 1960s, American Cold Warriors called on

Washington for intensified efforts to achieve US military superiority. Meanwhile, arms-control proponents, inside and outside of Washington, argued that negotiated limits on arms competition were more likely to lead to long-term security than both sides scrambling to gain a temporary military edge. 'The problem posed to both sides by this dilemma of steadily increasing military power and steadily decreasing national security', physicist and diplomat Herbert York insisted, 'has no technical solution'. Political solutions were needed.

In his 1969 inaugural address, Richard Nixon spoke of 'a new era of negotiation' in which all nations, especially the superpowers, would seek 'to reduce the burden of arms' while reinvigorating the 'structure of peace'. This could be accomplished, Nixon envisaged, through a programme of 'linkage', or détente. He and his national security adviser, Henry Kissinger, were prepared to go considerably beyond previous administrations in discussing strategic arms control and trade issues with the Soviet Union, but they expected the Kremlin to reciprocate by assisting in the resolution of ongoing disputes in Africa, the Middle East, and South-East Asia.

In November 1969, the superpowers' delegations began bilateral talks focused on the limitations of both defensive and offensive strategic weapons systems—essentially ICBMs and SLBMs. These negotiations would continue, intermittently, resulting in two strategic arms limitations treaties (SALT I and II), the Intermediate-Range Nuclear Forces (INF) Treaty (the only treaty that actually reduced the number of offensive nuclear weapons during the Cold War), and the strategic arms-reduction talks (START I) that were finally concluded in 1991.

The SALT I pacts of May 1972 consisted of the Anti-Ballistic Missile (ABM) Treaty which limited each party to two sites, an Interim Agreement (1972–7) on strategic systems, and a political 'Basic Principles' accord. The Interim Agreement's limits on strategic systems were actually higher than each currently possessed; but it did set ceilings on future deployments. To defeat Soviet ABM systems, in 1967 the US began developing a multiple independently targeted re-entry vehicle (MIRV), which carried aloft on a single missile several warheads, each capable of striking a different target. Delegates might have halted MIRV programmes during SALT I negotiations, but Pentagon and congressional opponents had warned Kissinger, 'don't come back with an MIRV ban'. Three years later, when Moscow deployed its own, considerable MIRVs, the Pentagon paid for its short-sighted insistence on a temporary advantage as the MIRV deployments made a pre-emptive strike appear more promising in a crisis situation because each side's ICBMs had become vulnerable.

The 'Basic Principles of Relations' agreement was initiated by the Kremlin and, while generally ignored by the American leadership, was considered by Soviet officials as 'an important political declaration'. They hoped it would be, as Dobrynin recalled, the basis of a 'new political process of détente in our relations' because it recognized the Soviet doctrine of peaceful coexistence and acknowledged the 'principle of equality as a basis for the security of both countries'. Moscow believed the superpowers could cooperate in resolving their basic differences despite 'minor' problems in the Third World; however, Washington interpreted détente to mean that the USSR, China, and Cuba were to maintain a 'hands-off' policy in the Third World. The failure to develop détente's boundaries and to gain public acceptance doomed the idea. American hawks vigorously denounced any attempt to ameliorate relations with the Soviet Union.

President Gerald Ford and Soviet Premier Leonid Brezhnev agreed 'in principle' at Vladivostok in November 1974 that each side should be limited to 2,400 ICBMs, SLBMs, and long-range bombers, of which 1,320 could have multiple warheads; but they could not finalize a SALT II pact. After stumbling in his initial efforts, President Jimmy Carter finally agreed to a seventy-eight-page SALT II treaty in April 1979 that hewed closely to the so-called Vladivostok principles but also limited air-to-surface cruise missiles and carried an extensive list of qualitative restrictions. He failed, however, to press for its ratification.

Ronald Reagan had never, until he met Soviet leader Mikhail Gorbachev, supported an arms-control treaty. He opposed the 1963 Limited Test Ban pact, the 1968 Non-Proliferation Treaty, and the 1972 SALT I and ABM agreements, and denounced SALT II as 'fatally flawed'. Moreover, early in Reagan's first term he ended negotiations for a comprehensive test ban treaty, and terminated US compliance with SALT II in May 1986. Contrary to Reagan apologists, Gorbachev's concessions were essential for the arms-control accomplishments during the Reagan presidency.

In May 1982, Reagan announced a plan for a 'practical phased reduction' of strategic weaponry. If the public was enthusiastic, analysts labelled the initial START I plan non-negotiable because it required the Soviets to dismantle their best strategic weapons, while the US kept most of its Minutemen missiles, deployed one hundred of the new large MX (Missile Experimental) missiles, deployed its new cruise missiles, and modernized its submarine and bomber fleets. Attempts to modify the plan during the next four years met with interminable bickering between government agencies, prompting a senior member of the National Security Council to acknowledge that 'Even if the Soviets did not exist, we might not get a START treaty because of disagreements on our side.' Another high-ranking US official complained that if the Soviets 'came to us and said, "You write it, we'll sign it," we still couldn't do it'.

As he began preparing for re-election in January 1984, President Reagan faced a multifaceted dilemma: how to ease tensions with Moscow, deflect the criticism of the anti-nuclear protestors both at home and abroad, and appease the hardliners in the Senate eager to chastise the Soviets for alleged arms-control violations. William Casey, director of central intelligence, had advised Reagan that NATO's exercise ABLE ARCHER that simulated nuclear response procedures had alarmed Soviet intelligence officials, who thought it might be a prelude to a nuclear attack. The president could not believe that Moscow might be genuinely fearful of an American attack, but on 16 January he spoke of 'reducing the risk of war, and especially nuclear war', through arms control, while raising questions about Soviet compliance and possible evasions of previous treaties. Reagan's peace appeal to the Russians, followed by charges of Soviet cheating, provided the former governor of California with trump cards for the 1984 campaign that the Democrats found hard to top.

Subsequently, a series of reports to Congress claimed a variety of Soviet violations (and the Soviets responded with their own list of US evasions), most of which were 'grey area' complaints. Moscow was guilty, however, of two significant violations—an uncompleted radar site, violating ABM terms, and a vast experimental biological warfare project (largely undiscovered until after the Cold War) violating the Biological Warfare Convention.

At the Reykjavik summit in October 1986, Reagan suggested the elimination of all ballistic missiles within ten years. Secretary General Gorbachev immediately countered with the elimination of all Soviet and US strategic nuclear weapons within ten years and restricting the Strategic Defense Initiative—Reagan's missile defence scheme, dubbed 'Star Wars' by the media (see Chapter 6)—to an experimental stage for a decade. When Reagan refused to accept any limitations on his 'Star Wars' project, these radical arms-reduction proposals were dropped—much to the relief of US military leaders and NATO members and, undoubtedly, to senior Soviet marshals.

Nonetheless, there was a significant breakthrough at Reykjavik when Gorbachev agreed to American demands for on-site inspections. With the Limited Nuclear Test Ban and the SALT I pacts, Washington had settled for verification by national technical means—employing satellite reconnaissance, electronic monitoring, and other self-managed intelligence-gathering techniques. After Reykjavik, it was the Soviets who insisted on intrusive inspections, but the Pentagon and intelligence agencies began having second thoughts as they realized they did not want the Soviets prowling US defence plants. As Secretary of Defense Frank Carlucci admitted, 'verification has proven to be more complex than we thought it would be. The flip side of the coin is its application to us. The more we think about it, the more difficult it becomes.'

Shortly after Reykjavik, Gorbachev again surprised NATO and Washington leaders by accepting the US's 'zero-option' for an intermediate-range missile pact that required disproportionate Soviet reductions, including their missiles in Asia. On 8 December 1987, he and Reagan signed the INF Treaty that included the first nuclear arms reductions and an elaborate US–Soviet on-site inspection system. The INF Treaty, a central pillar of the international arms-control regime set in place during the last stage of the Cold War, would last three decades. Not until after the conclusion of the Cold War did Presidents George H. W. Bush and Gorbachev sign the complex 750-page START I treaty, in July 1991. This was the first agreement to call for significant cuts in strategic weaponry, as almost 50 per cent of nuclear warheads carried on each power's ballistic missiles were to be eliminated. It was a very promising moment in the annals of limiting the bomb. It was also a fitting ending to the Cold War, which ended with a whisper rather than mushroom clouds. Nuclear stability prevailed due to good luck and mutual prudence.

Chapter 6
Star Wars and beyond

At the onset of the Cold War in the late 1940s, American officials believed that the United States' sole possession of atomic bombs would simply deter the Soviet Union from expanding further into Western Europe or Asia. After the Soviets developed atomic weapons and aircraft capable of delivering them over the North Pole during the early 1950s, the United States accelerated its efforts to obtain missiles capable of shooting down enemy bombers. The advent of thermonuclear warheads by the late 1950s and the deployment of nuclear-tipped ICBMs by the early 1960s, spurred the search by both superpowers for viable anti-ballistic missile defence systems (ABM systems).

Both Washington and Moscow found themselves caught up in an offensive and defensive arms race that threatened the stability of the embryonic nuclear deterrence system. As the concept of deterrence began to take hold, initial concerns arose as to whether an ABM defence system would actually provide much 'defence', and whether it was cost-effective. Eventually US domestic politics, driven by partisanship and the threat of 'the axis of evil', overcame previous concerns about costs and effectiveness as President George W. Bush ordered the deployment of an untested ABM system in 2002.

Initial US missile defence projects

The United States' defensive missile programmes began in November 1944 when the US army contracted with the General Electric Company to investigate ways to protect American forces from Germany's V-2 rockets. Later, General Electric's research on ballistic missile defences was accelerated with the assistance of captured German documents and German scientists arriving in 1946. Within twelve months it had assembled and fired one hundred V-2 missiles to obtain essential data about an offensive ballistic missile's trajectory and re-entry into the atmosphere. Research eventually led to the Nike-Ajax, the army anti-aircraft missile, in 1953, and Nike-Hercules, the army's follow-on anti-aircraft missile system, the next year.

Two Soviet developments in 1957 alarmed Americans and, at the same time, challenged their scientists to develop an anti-missile system. In August, the Soviets tested an intercontinental ballistic missile, and in October they startled the world by launching Sputnik I, the first orbiting artificial satellite. These events raised questions about the United States' vulnerability to a surprise nuclear attack—an impression that Soviet leaders were keen to foster, as they announced its rockets were able to reach any part of the globe. President Dwight Eisenhower created a high-level commission, led by Rowan Gaither, which recommended, among other things, the development of an ABM defence system that would protect Strategic Air Command's missile bases.

Domestic politics and the desire to stabilize the nuclear environment played a major role in American and Soviet ABM decisions after the 1962 Cuban Missile Crisis. Members of Congress, alarmed at the United States' vulnerability during the crisis, urged the president to immediately deploy a national ABM system. The Soviets, meanwhile, upgraded their liquid-fuelled long-range missiles—which took considerable care and time to prepare for launch—with more dependable, quickly launched solid-fuel ICBMs. By 1967, the Soviets had an estimated 470 solid-fuel ICBMs, while the United States possessed 1,146, suggesting both superpowers possessed more than enough missiles to effectively deter each other. Unless, perhaps, one side possessed an effective national ballistic missile defence system.

President Lyndon Johnson's 24 January 1967 budget message to Congress indicated that the development of the promising Nike-X ABM system would continue, but that it was not yet ready to be deployed. The Nike-X was basically an army ABM system, linking multiple-array radar with an interceptor missile. Subsequently, however, the chairman of the Joint Chiefs of Staff, General Earle Wheeler, told the House Appropriations Committee that the US should immediately deploy a light missile defence, but acknowledged that the Joint Chiefs preferred a heavy ABM city defence system for 'the highest density populated areas'. Wheeler insisted 'the Nike-X was ready for deployment'. Other prominent Americans, including the Committee for a Prudent Defense Policy, wanted a broad-based US ABM system deployed to meet the Soviet Galosh ABM system's challenge to stability of deterrence.

Despite considerable pressure, Secretary of Defense Robert McNamara questioned the effectiveness of the Nike-X system and worried that ABM systems were becoming a destabilizing factor endangering the existing nuclear parity between the US and the Soviet Union. He urged President Johnson to go slow because there were two other, more cost-effective alternatives: (1) improvement of the United States' offensive capabilities; and (2) consultation with the Soviets about the possibility of limiting offensive and defensive strategic arms.

At the brief June 1967 summit meeting with President Lyndon Johnson at Glassboro, New Jersey, Premier Aleksei Kosygin insisted that the Soviet Union's projected defensive missile systems 'don't kill people. They protect them.' Moreover, he insisted, 'Defense is moral; offense is immoral.' Ironically, three and a half decades later, James M. Lindsay and Michael E. O'Hanlon argued that 'a national security policy that deliberately leaves the American people vulnerable to attack when technology makes it possible to protect them is immoral and unacceptable. Not only does it fly in the face of common sense to leave the nation undefended, but it could hamstring America's role in the world.'

Without defences, proponents of ABMs believed that governments hostile to the United States that possessed nuclear-tipped ballistic missiles might well believe they could threaten America's extensive worldwide interests and thus deter Washington from taking measures to protect them. Also, without an adequate missile defence the United States' allies might question Washington's willingness to honour its security pledges and thus lessen its global influence. Later, fears arose in the US that terrorist groups might obtain ballistic missiles with nuclear warheads to target American cities.

In contrast, opponents of ABM programmes questioned the high costs and effectiveness of projected US ballistic missile defences. They have also worried about the destabilizing impact of such anti-missile systems on relations with allies and adversaries. Would rival nations fear that the United States—should Washington believe the US to be impervious to retaliation—might flaunt its strategic arsenal as a means of pressuring them to conform to Washington's wishes or face serious consequences? Would US missile defences cause a fearful opponent to feel compelled to strike first, early in a crisis, with full force? Would such activity, in fact, impede strategic arms-limitation efforts? Would the next step be to place nuclear weapons in space? Would US missile defences renew the strategic nuclear arms race?

Thus, opponents contended that should a nationwide missile defence system result in an enemy considering launching a first strike, in stimulating an arms race in outer space, or in the proliferation of ballistic missiles and weapons of mass destruction, Americans would find themselves with much reduced security. They repeatedly urged that strategic arms-control activities not be sacrificed in a dubious, costly quest for technological solutions or squandered in unilateral ventures.

Soviet missile defence projects

The United States' monopoly of nuclear weapons in the late 1940s, and the possession of bombers to deliver them, prompted the Soviet Union to concentrate on defensive systems. In 1947, the Soviets began experimenting with anti-aircraft missiles modelled on Germany's World War II rockets, and eventually, on 25 May 1953, their V-300 missile and radar guidance system successfully shot down a TU-4 unmanned bomber. Six months later, the construction of an anti-aircraft missile defence system (S-5) began around Moscow to shield the city from up to 1,000 attacking bombers; in 1956 the defensive ring was designated to receive the Soviet's first ABM system (A-35 or 'Galosh') by November 1967. However, tests of its new S-350 interceptor missile indicated it could not cope with the US's new MIRV. Each US ICBM re-entry vehicle (often referred to as a bus) could now carry several decoys and three or more individual nuclear-armed warheads.

Meanwhile, the Soviets decided in 1974 to develop the A-135 ABM system as a replacement for the A-35. The A-135 had been designed to counter either single or MIRVed ICBMs and was to have a two-tier defence capability. The first tier of interceptor missiles with A-350 launchers would attack ICBMs outside the atmosphere (exoatmospheric), while the second tier of A-350 launchers would deal with ICBMs in the atmosphere (endoatmospheric). The first-tier system was confronted with the difficulty of locating and differentiating between decoys and warheads, the most serious problems confronting any ABM system. Following successful tests of the two-tier system at Sary Shagan in 1975 and 1976, the Minister of Defence authorized construction of seven A-135 sites around Moscow, beginning with the multipurpose Don-2N radar system in 1978, and continuing with hardened missile silos in 1981 that were completed in November 1987. However, the A-135 system did not become fully operational until around 1997.

The Russians still have little confidence in the ability of their ABM systems to stop the penetration of ballistic missiles. Consequently, since the end of the Cold War they have concentrated on improving their ICBMs and equipping them with decoys to defeat any US ABM system.

The United States' unrestricted development of nuclear-tipped cruise missiles, which could be launched from bombers or submarines, confronted the Soviet Union with new threats. After launching, American cruise missiles could fly at low altitudes, enabling them to enter Soviet territory without being detected by the Soviets' existing radar, and allowing them to penetrate deep into Soviet territory to destroy ICBMs in their silos.

To protect their ICBM silos and administrative and industrial sectors from cruise missiles, Soviet scientists between 1975 and 1980 sought to develop a theatre defence system employing a standardized multi-channel surface-to-air missile—the SAM-300 system. The S-300V could protect the Soviet army's ground units, the S-300F defended naval ships, and the S-300P protected air defence forces. The S-300P systems had their equipment and launchers mounted on mobile trailer platforms connected by cables and given the name S-300PT. In 1980, the S-300PT system using the 5V55 surface-to-air missile was deployed around Moscow to supplement the A-135 system. The S-300PT system remained on station until 1985, when it was replaced by an upgraded SS-300PM mounted on self-propelled trailers designed to traverse almost any terrain and linked by radio-relays to command and control centres.

In 2005–6, the Russian air force began replacing its S-300P with the S-400 (NATO reporting name SA-20 Triumf) surface-to-air missile systems mounting an upgraded 48N6DM long-range interceptor designed to destroy aircraft, cruise missiles, and short- and medium-range ballistic missiles at ranges of up to 400 kilometres (250 miles). The S-400 has approximately 2.5 times the range of the S-300P, and twice the range of the US Patriot Advanced Capability-3 (PAC-3) system. Lightweight 9M96 interceptor missiles, with a range of about 120 kilometres (75 miles), will be mounted to counter low-flying targets. As *Jane's Missiles and Rockets* subsequently reported, eventually all thirty-five regiments will be equipped with the new system, which will be used to protect large population centres, as well as military and industrial complexes.

Moscow has been aggressively marketing the S-400 throughout Asia, Europe, and the Middle East. Between 2003 and 2004, China spent approximately $500 million on future S-400 systems. Additionally, Russia has offered the S-400 to the United Arab Emirates, and there is also speculation that Iran, a potential nuclear power, is currently seeking to acquire its own S-400 missiles. Once the S-400 completed its final tests and entered production, it was expected to become one of the most sought-after missile defence systems in the world. Yet as the US Patriot systems proved in two Gulf wars, the American and Russian systems still had deficiencies to deal with and could not guarantee that an enemy's cruise or short-range missile would be stopped.

First US ABM deployment

In September 1967, the beleaguered Johnson administration agreed to deploy a 'thin-line' Nike-X ABM system to protect the US from China's less potent nuclear missile threat, but made it clear the proposed ABM system would not effectively protect the

US from a Soviet ICBM attack. By targeting China, the proposed Sentinel system left the door open for the Soviet Union to consider seriously the limitation or reduction of ABMs and ICBMs. It fell to the Nixon administration to undertake the actual deployment.

Shortly after his inauguration, Nixon announced on 14 March 1969 that: 'After a long study of all of the options available, I have concluded that the Sentinel program previously adopted should be substantially modified.' The new ABM system would 'not provide defense for our cities' because 'I found that there is no way that we can adequately defend our cities without an unacceptable loss of life.' Therefore, in 1970, he authorized the new Safeguard system to protect up to twelve Minuteman III ICBM sites at Malstrom AFB, Montana, and at Grand Forks AFB, North Dakota, in order to preserve a credible deterrent.

Nixon chose not to mention the enhancement of the Safeguard ABM system, which increased the number of ABM interceptors to protect Minuteman III ICBM sites and altered Sentinel's radar range to cover the continental United States. Kissinger's memoirs indicated that the extended radar coverage would create 'a better base for rapid expansion' of ICBM site defences if needed in the future. (Soviet scientists correctly anticipated that omitting data about radar coverage was part of the Safeguard plan.)

Because of Safeguard's technical limitations, the House of Representatives voted on 2 October 1975 to deactivate the single ABM site (instead of the planned twelve) at Grand Forks, North Dakota, after spending $6 billion—some four months after it became operational. This action followed the realization that Safeguard's large phased-array radars provided easy targets for Soviet missiles and, additionally, that when nuclear warheads on the Spartan and Sprint missiles detonated, the explosions blinded the radar system.

The 1972 ABM Treaty

Meanwhile, the first steps towards US–Soviet negotiations on missile defence systems began in 1964 when William Foster, US director of the Arms Control and Disarmament Agency, explored the possibility of negotiations to ban or place limits on the ABM systems with Anatoly Dobrynin, the Soviet Union's ambassador to the United States. Moscow did not act on these initial American suggestions, according to Dobrynin, because members of the Politburo could not agree on whether to negotiate with Washington. On 10 August 1968, the Kremlin finally decided to begin discussions to limit or eliminate offensive and defensive strategic weapons. Unfortunately, these planned talks were sidetracked on 24 August, when Soviet forces intervened in Czechoslovakia.

Strategic arms discussions that began in Helsinki, Finland, on 17 November 1969, found American delegates pressing their concern, both formally and privately, that ballistic missile defence systems endangered current deterrence stability. Early in the talks, the Soviets indicated a willingness to limit ABM deployment 'to geographically and numerically low limits'. Confronted with unresolved issues regarding strategic offensive forces, it was finally agreed in 1971 to seek separate agreements.

During late August 1971, American delegate Harold Brown was asked to clarify the United States' 'understanding of the notion of "development" and of practical application of limitations'. After checking with his superiors, Brown carefully responded that:

> By 'development' we have in mind that stage in the evolution of a weapon system which follows research (in research we include the activities of conceptual design and laboratory testing) and which precedes full-scale testing. The development stage, though often overlapping with research, is usually associated with the

construction and testing of one or more prototypes of the weapon system or its major components. In our view, it is entirely logical and practical to prohibit the development—in this sense—of those systems whose testing and deployment are prohibited.

Unknowingly, Brown had provided a definition that would be employed in the 1980s by opponents of President Ronald Reagan's 'Star Wars' system to reinterpret the 1972 ABM pact. By the autumn of 1971, the Soviet and American delegates, meeting at Geneva, agreed on the basic elements of Article V of the ABM Treaty that read: 'Each Party undertakes not to develop, test, or deploy ABM systems or components which are sea-based, air-based, space-based, or mobile land-based.' Fixed land-based systems were defined in Article II as 'a system to counter strategic ballistic missiles or their elements in flight trajectory, currently consisting of' ABM interceptors, launchers, and radars. The phrase 'currently consisting of' indicated that the treaty was to cover all systems—current and future.

The Soviets exercised a persistent inquisitiveness regarding 'exotic' systems, partly because for months American delegates were prevented—by their military chiefs—from using lasers as an example. The Soviets were aware of America's laser programme and, in fact, hoped to employ their own large lasers in anti-missile experiments. Eventually, Soviet probing and, perhaps, their hope to glean information about the US exotic programme, gave way to an agreement to ban the deployment of fixed-based exotic ABMs. In the Agreed Statement D of the ABM Treaty, a footnote stated that:

> the Parties agree that in the event ABM systems based on other physical principles and including components capable of substituting for ABM interceptor missiles, ABM launchers, or ABM radars are created in the future, specific limitations on such systems and their components would be subject to discussion . . . and agreement.

The ABM Treaty limited each side to two ABM sites (later reduced to one) separated by no less than 1,300 kilometres (800 miles), to keep them from overlapping. Consequently, each of the two permitted sites was restricted to specific areas and could only provide limited coverage. The treaty clearly prohibited the establishment of a nationwide ballistic missile defence system. At Moscow, on 22 May 1972, the terms of the ABM Treaty agreement were finalized and signed.

Reagan's Star Wars proposal

Following a Pentagon's Defense Science Board review, the White House concluded in October 1981 that its 'ballistic missile technology [was] not at the stage' where it can provide 'defenses against Soviet missiles'. This finding, according to Ronald Reagan's biographer, Lou Cannon, did not lessen the president's 'vision of nuclear apocalypse and his deeply rooted conviction that the weapons that could cause this hell on earth should be abolished'. Moreover, Reagan was morally opposed to the US's twenty-year-old deterrence doctrine, 'assured destruction'.

In early 1983, President Reagan was preparing a speech in support of another increase in the Defense Department's budget for fiscal year 1984 that was being challenged by a grass-roots nuclear freeze movement. He rejected the first draft because it repeated previous justifications. Rather than rehash old themes, Reagan urged his national security adviser, Robert C. McFarlane, to develop something new to counteract the message of nuclear freeze proponents. Public opinion polls in 1982 and January 1983 revealed that 66 per cent of Americans believed Reagan was not performing well in promoting arms control, and 70 per cent supported a freeze on nuclear weapons production as a first step to eliminating all nuclear warheads. A congressional debate on a nuclear freeze, which threatened increases in military expenditures, was scheduled for the end of March 1983.

Senator Malcolm Wallop (R-WY), Lt General Daniel O. Graham (retired), and physicist Edward Teller of the University of California Lawrence Livermore Laboratories had been lobbying the Pentagon and Congress from 1979 to 1982 for increased funding of missile defence projects. They sought support for such concepts as nuclear and chemically based lasers, orbiting space-based battle stations using lasers, and an improved air force space-aircraft. In February 1981, Secretary of Defense Caspar Weinberger told the Senate Committee on Armed Forces that the US might be able to 'deploy MX [missiles] in fixed silos protected by ABMs'. However, none of these advocates of anti-missile systems was directly involved in preparing Reagan's March 1983 speech.

On 11 February 1983, Reagan and the Joint Chiefs of Staff discussed the Pentagon's list of five options to deal with current strategic weaponry. One option was chief of naval operations, Admiral James Watkins's proposed missile defence system. He argued that a forward strategic ballistic missile defence would 'move battles from our shores and skies'. Such battles would be 'moral' and palatable to the American people because a missile defence system would protect Americans, 'not just avenge them', after a Soviet attack. It seemed realistic, Watkins concluded, to have a long-range programme to 'develop systems that would defeat a missile attack'. Reagan gravitated to Watkins's missile defence idea as a way to alleviate his moral aversion to the reality of nuclear deterrence.

Meanwhile, McFarlane and the president's science adviser, George Keyworth II, were drafting Reagan's speech scheduled for 23 March. Keyworth initially opposed inclusion of a missile defence plan, but reluctantly withdrew his objections after McFarlane informed him that inclusion of the proposed missile defence system was a political, not a scientific, decision.

According to Reagan's autobiography, he received a final draft of the speech on 22 March and that night 'did a lot of rewriting.

Much of it was to change bureaucratese [*sic*] into people talk.' In its finished form, his speech began with a lengthy section designed to persuade Congress to approve a significant increase in funds for the 1984 fiscal year to continue the US military build-up. As his speech drew to a close, Reagan told his audience of recent discussions about missile defence with the Joint Chiefs of Staff. Then, after noting that the nation's security previously depended on nuclear deterrence, Reagan continued:

> Let me share with you a vision of the future which offers hope. It is that we embark on a program to counter the awesome Soviet military threat with measures that are defensive.... What if free people could live secure in the knowledge that their security did not rest on the threat of instant U.S. retaliation to deter a Soviet attack, that we could intercept and destroy strategic ballistic missiles before they reached our own soil or that of our allies?

Acknowledging this would be a formidable undertaking, he suggested that as current technology offered promise, it was time to begin creating a defensive shield:

> I call upon the scientific community in this country, who gave us nuclear weapons... to give us the means of rendering these weapons impotent and obsolete. Tonight, consistent with our obligations of the ABM Treaty... I'm taking an important first step. I am directing a comprehensive and intensive effort to define a long-term research and development program to begin to achieve our ultimate goal of eliminating the threat posed by strategic nuclear missiles.

The president's proposal was officially titled the Strategic Defense Initiative (SDI) in January 1984, while critics dubbed it 'Star Wars'.

The response to Reagan's proposal was decidedly mixed. Undersecretary of Defense Richard DeLauer, who endorsed funding ABM research, objected to it being subjected to such a 'half-baked political travesty'. When cornered by a reporter, Minority Whip Robert Michel of Illinois said the speech may have been 'a bit of overkill'. *Time* magazine's lead story after the speech suggested Reagan's proposal was representative of a 'video-game vision', and its cover pictured Reagan against a background of space weapons resembling a Buck Rogers comic strip about the 25th century. Within a week, however, Reagan's missile defence proposal had disappeared because it was no longer considered newsworthy, and the public's attention shifted to more immediate issues. Indeed, during the 1984 election campaign Reagan did not mention missile defence, even though Democratic candidate Walter Mondale denounced it as a dangerous hoax costing American taxpayers billions of dollars, speeding up the arms race while providing no real protection to the American people.

Taking the SDI proposal seriously, the Defense Department created two expert panels in the spring of 1983—the Fletcher and Hoffman groups—to examine possible missile defence systems. James C. Fletcher, former director of NASA, headed a sixty-five-member panel—fifty-three of whom had direct financial interests in SDI research—asked to plan a missile defence system. In early 1984, the panel recommended that all research aspects of SDI should be accelerated to reach a decision on deploying a missile defence system in the early 1990s.

The Fletcher panel proposed a layered-interceptor missile defence system. The first layer involved SDI sensors detecting ICBMs leaving their silos and the immediate launching of missile interceptors to attack the enemy missiles in their boost phase. The second layer of US interceptors would seek to destroy enemy warheads in the post-boost, or bus deployment, phase. The third layer of interceptors would look for any deployed enemy warheads during a midcourse phase before they entered the atmosphere. Finally, a fourth layer of interceptors would sort out surviving warheads from the decoys and debris during the terminal phase and destroy them (see Figure 7).

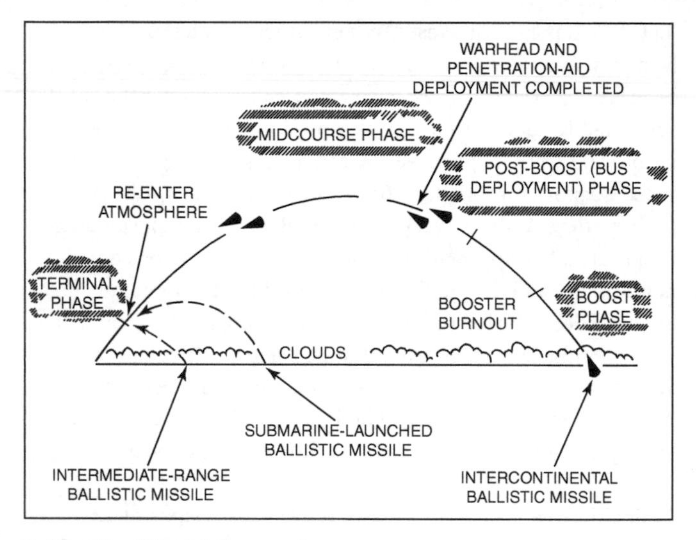

7. The 'Star Wars' defence system.

Destroying enemy ICBMs in the brief initial boost phase of three to five minutes would provide the best opportunity for reducing the number of incoming warheads. After the boost phase passed, the post-boost vehicle ('bus') would continue to carry the warheads and decoys. The post-boost phase would take six to ten minutes to reach its apogee of some 750 miles above the earth, during which time a second US layer of interceptors would try to find and destroy the bus. This is the next best time to intercept the nuclear warheads. At its apogee, the bus would adjust its trajectory and release up to ten nuclear warheads, plus numerous decoys, all of which would begin descending towards selected targets on earth.

The third layer of missile defence comes into play during the midcourse phase after the bus releases the warheads and decoys and before these objects re-enter the earth's atmosphere. This layer gives the US missile defence system its greatest amount of time, perhaps up to twenty minutes, to locate and destroy the incoming warheads that are heading towards their targets. US interceptors, however, may be diverted from the warheads by decoys and space debris that they could mistake for enemy warheads.

The final missile defence phase begins when warheads and decoys re-enter the atmosphere about sixty miles above the earth. During this phase, interceptors have only tens of seconds to hit warheads before they reach their target. The one advantage for defensive missiles at this stage is that the warhead's skin is heated by friction, while decoys, presumably more lightweight, cool down after they separate from the warheads.

For the ballistic missile defence system to qualify for deployment, it should effectively fulfil three tasks. First, the system must be able to detect and identify enemy targets, that is, distinguish among ICBM booster rockets, warheads, decoys, and debris. Second, the system's tracking devices must be able to locate and plot the trajectory of a target in order to guide an interceptor missile to its target. Finally, a defence system must be able to assess the damage caused by the defensive weapon to assure the destruction of the booster rocket, bus, or warhead. This is necessary so that defenders can determine whether they must launch additional interceptors.

Obviously, such a comprehensive ballistic missile defence posed a most daunting challenge to the scientists and technicians who were to undertake the necessary research to develop and test the complex parts of the system. It also required large increases in the Defense Department's budget, much larger than the estimates initially offered by the Reagan administration.

Meanwhile, arms negotiator and diplomat Paul Nitze had presented a three-part formula that any SDI system would need to meet before it could be considered for deployment. The 'Nitze criteria', as it was known, stated that the anti-missile system should: (1) be effective; (2) be able to survive against a direct attack; and (3) be cost-effective at the margin—that is, less costly to increase your defence than your opponent's costs to increase their offensive against it. Nitze's formula was adopted as National Security Directive No. 172 on 30 May 1985, prompting some at the Pentagon to fear that stressing cost-effectiveness would essentially kill the programme. Others, such as Robert McNamara, doubted that the Reagan administration planned to adhere to the cost portion of the criteria.

At the same time, a Future Security Strategy Study team—with seventeen of its twenty-four members future SDI contractors—chaired by Fred S. Hoffman, also assessed the nation's strategic defences. In early 1984, the Hoffman study offered a more realistic appraisal of the SDI's time frame. Rather than anticipating a possible ABM deployment in the early 1990s, Hoffman's team concluded that a perfect defence might 'take a long time and may prove to be unattainable in a practical sense against a Soviet effort to counter the defense'.

For more than a decade, the traditional interpretation of the ABM Treaty was seen as prohibiting any development and testing of a space-based ABM system. But in October 1985, Nitze persuaded Secretary of State George Schultz to accept a 'broad' interpretation of the 1972 Treaty that would permit R&D of space-based weapons. Other administration hardliners—who wanted to scrap the ABM Treaty—sought instead to broaden the new interpretation further to permit testing of space weapons.

On 6 October, National Security Adviser Robert McFarlane told NBC's 'Meet the Press' that the 1972 ABM Treaty allowed development and research of a missile defence system that involved 'new physical concepts'. He also argued that the treaty permitted the testing of exotic systems and technologies—presumably lasers and particle beams.

The State Department's legal adviser, Abraham D. Sofaer, argued that the classified ABM Treaty negotiation record and treaty provisions showed its language to be ambiguous and that the record of Senate ratification supported the broader view. He also claimed, without providing any substantiation, that the Soviet Union never accepted a ban on mobile ABM systems or on exotic technologies. (Sofaer eventually had to acknowledge that the ratification records did not support the broad interpretation and blamed the errors on his staff's 'young lawyers'.)

The administration's efforts to broaden the interpretation of the 1972 ABM Treaty provoked a major executive–legislative disagreement. Warning the president that any actions in violation of the traditional interpretation of the pact would cause 'a constitutional confrontation of profound dimensions', Senator Sam Nunn (D-GA) launched a series of studies of the reinterpretation, concluding that Sofaer's legal reasoning was in 'serious error'. Joined by Senator Carl Levin (D-MI), Nunn sponsored an amendment to the defence authorization bill prohibiting any SDI testing that challenged the traditional interpretation of the ABM Treaty prohibitions. Following a sharp partisan debate and an extended Republican filibuster, a modified version of the Nunn–Levin wording was approved in late 1987.

The Republicans won control of the House and Senate in 1994 and attributed victory to their 'Contract with America' that among other issues reflected how deeply the commitment to a nationwide missile defence had become enmeshed in the party's political ideology. It called for deploying a 'cost-effective, operational anti-ballistic missile defence system' as early as possible to protect the United States 'against ballistic missile threats (for example, accidental or unauthorized launches or Third World attacks)'. Moreover, the contract insisted that the ABM Treaty was 'a Cold War relic that does not meet the future defence needs of the United States.... It is a moral imperative that US strategic defenses be expanded and that the Clinton administration not yield to Russian demands that Americans remain defenseless in the face of potential nuclear aggression.' During subsequent years, Republican legislators unsuccessfully sought to mandate deployment of a national missile defence system.

The Republicans appointed an independent commission in November 1996 to 'Assess the Ballistic Missile Threat'. Under the direction of future Secretary of Defense Donald Rumsfeld, the Rumsfeld commission's declassified Executive Summary emphasized that: 'The newer ballistic missile-equipped nations [North Korea, Iran, and Iraq]...would be able to inflict major destruction on the US within about five years of a decision to acquire such a capability (10 years in the case of Iraq).' North Korea and Iran, who the commission thought to be developing weapons of mass destruction, were alleged to have put 'a high priority on threatening US territory, and each is even now pursuing advanced ballistic missile capabilities to pose a direct threat to US territory'.

Greg Thielmann, formerly of the State Department's Bureau of Intelligence and Research, found 'Rumsfeld's view on ballistic missiles often ignored the carefully considered views of [intelligence] professionals in favor of highly unlikely worst-case scenarios that posited an imminent threat to the United States and prompted a military, rather than diplomatic, response.' This was not surprising.

George W. Bush and ABM deployment

Not long after being inaugurated as president in January 2001, George W. Bush undertook to fulfil his campaign promise to actively pursue a national missile defence system. After the 11 September 2001 terrorist attacks, Bush insisted that a missile defence system was necessary for American security. To remove all limitations on research, development, and testing of missile defences, he announced on 13 December 2001 that the United

States had given Moscow the required six months' notice of its intention to withdraw from the 1972 ABM Treaty.

A year later, in December 2002, President Bush instructed the Defense Department to deploy the initial elements of a strategic missile defence system. The modest deployment included twenty ground-based midcourse missile defence (GMD) interceptors and twenty sea-based Aegis ballistic missile defence interceptors positioned on three vessels. Also included were an unspecified quantity of Patriot PAC-3 missiles and upgraded radar systems to help locate potential targets. The PAC-3 missiles and the sea-based interceptors were intended to protect against short- and medium-range ballistic missiles. Only the twenty GMD interceptors—sixteen to be placed in Alaska and four located at Vandenberg Air Force Base—were designed to protect against long-range ballistic missiles. Informed observers fully understood that the intercept tests of the rudimentary GMD intercepting rockets had been carefully scripted with modest challenges; even the 'successful' ones did not resemble real-world conditions. A reliable ABM system appeared still to be years away.

Further considerations

Among several remaining contentious considerations are three questions that deserve further comment: (1) Will a missile defence system provide the best defence against rogue states and terrorists? (2) Has the political partisanship that drove the deployment decision become a faith-based commitment? (3) What will the missile defences cost? (See Box 3.)

Most Americans agree that possession of an effective missile defence system would be desirable; however, many sceptics are concerned that the precipitous deployment of unproven systems, at substantial expense, could fall far short of providing the desired shield. Several analysts have contended that given the United States' enormous nuclear arsenal and global delivery capabilities,

Box 3 Early US missile defence systems (a selective list of American missile programmes)

Project	Date	Research and purpose
Thumper	1944	Army research seeks protection from V-2 type rockets, leads to BAMBI (Ballistic Missile Boost Intercept), cancelled in 1961
Nike	1945	Army launches research for anti-aircraft defence
Gapa	1947	Air force seeks Ground-to-Air Pilotless Aircraft, integrated in 1949 with Thumper to hit and kill ballistic missiles
Bumblebee	1947	Navy seeks surface-to-air missile, leads to Talos
Nike-Ajax	1953	Army anti-aircraft missile
Nike-Hercules	1954	Army anti-aircraft system
Wizard	1955	Air force's ABM, eventually shifts to offensive missiles
Nike-Zeus	1956	Army ABM system, links radar with interceptor rocket
Talos	1958	Eventually becomes Polaris SLBM
Nike-Zeus	1960	Army urges deployment to protect military bases, leads to Nike-X
Nike-X	1963	Multiple-array radar and Sprint missile added to system
Sentinel	1968	Designated as Sentinel, Nike-X to be deployed nationwide against China
Sentinel	1969	Becomes Safeguard and to be deployed at North Dakota/Montana ICBM silos
Safeguard	1975	Safeguard becomes operational
Safeguard	1976	Congress orders Safeguard shut down

Soviet urban air defence systems		
Moscow		
A-25	1953	Anti-bomber defence uses V-300 surface-to-air missile
A-35	1958	Construction begins on GALOSH* system—planned to protect from ICBMs by 1967 using V-1000 missile
	1962	S-350 interceptor added to operate outside the atmosphere but it fails to counteract MIRVs
	1967	Work stopped on GALOSH due to ineffective testing and Moscow is defended only by Aldan system of TU-126 fighter aircraft
	1975	A-350 interceptor upgrades against MIRVs
A-135	1978	System upgraded gradually
	1980	5V55 provide protection for air defence units
	1992	Replaces A-35
Leningrad		
	1961	Uses S-500 (Griffin*) interceptors with single-stage SAM launcher—abandoned 1963
	1963	S-200 (Gammon*) interceptor with two-stage SAM launcher
	1970	S-200V Volga increases range and adds ABM capabilities
	1974	S-2S-00D Vega—upgrade of S-200V is abandoned after amended 1972 ABM Treaty limits each party to one site

* US designation

no nation would allow the launch of a ballistic missile from its territory because such hostile action would result in immediate American retaliation and annihilation of the offending state.

A much more likely threat to the United States, according to these specialists, is that foreign terrorists, if they chose to use weapons of mass destruction, would employ a ship or truck to carry them to American soil—not long-range ballistic missiles which are complicated to build, deploy, and launch with accuracy. Thus, America's greatest threat, in the words of one commentator, is not from rogue states, but from stateless rogues.

Ever since President Reagan's SDI speech, the heated debates and demands for immediate deployment of a missile shield can be traced to the domestic political environment. So-called 'conservative' Republicans became increasingly strident in their determination to terminate the 1972 ABM Treaty and to deploy an anti-missile system. This commitment bordered on theological cant, appeared in official party documents, and brooked little or no compromise. Moreover, Republican demand for deployment paid little attention to time-proven procedures for developing weapons systems, to concerns of various technological deficiencies of the anti-missile systems, to the financial costs, or to the impact of deployment upon broader foreign policy considerations.

A third consideration is that past R&D activities have consumed more than $120 billion, and costs will continue to mount with the decision to deploy unproven technology. The head of the Missile Defense Agency (MDA), General Ronald Kadish, illustrated the administration's lack of concern regarding the costs of what many regard as premature deployment, because he proposed to 'Test, fix. Test, fix. Test, fix.' While this is the usual process in the experimental stage, it becomes more expensive once 'operational' units are fielded.

In its June 2003 report, the General Accounting Office (GAO) questioned the wisdom of the Pentagon's push for deploying a limited missile defence at the expense of ignoring the proven approach to developing weapons systems and for employing a 'test, fix' policy. Consequently, the GAO warned that the administration was risking the deployment of costly, ineffective anti-missile systems.

The MDA has estimated that the deployment would probably cost an additional $50 billion. The GAO, however, emphasized that this sum only related to R&D expenses. It did not include the cost of production, operations, and maintenance that earlier Pentagon figures estimated could be nearly another $150 billion. The GAO urged the Pentagon to consider preparing a comprehensive estimate of missile defence costs and that it should begin budgeting for these expenditures. Failure to do so could result in the Defense Department being forced to shift funds from other weapons programmes to meet the costs of building and deploying the missile defence system.

Forecasting the costs of a layered-missile defence system is quite daunting. However, when the Economists Allied for Arms Reduction added up the Pentagon's own estimates for all of the elements of the various phases of the Bush administration's projects, including operating the systems for twenty years, they found that it totalled a trillion dollars, maybe a trillion and a half. In a world in which the global strategic nuclear environment was rapidly changing, even that was not enough.

Chapter 7
Post-Cold War era

When George H. W. Bush assumed the presidency in January 1989, it was generally expected that he would expedite the arms-limitation process begun by Ronald Reagan and Mikhail Gorbachev. Clearly, that did not happen. Because Bush believed Reagan had casually granted too many concessions to Gorbachev, he and his National Security Adviser Brent Scowcroft slowed down the momentum of US–Soviet negotiations, emphasizing the continued need of vigilance and strength in dealing with the Kremlin. Neither the new president nor many of his senior advisers truly believed the Cold War was over.

Bush and his secretary of state, James Baker, believed that new presidents in the past had gotten into trouble by moving too quickly; thus, it would be better to get a solid grasp of the status of Soviet–American relations before opening formal discussions with Moscow. Scowcroft was even more cautious as he advised the country in January that the West should keep up its guard, because Gorbachev could well be a new version of the 'clever bear syndrome'—seeking to lull the West into a false sense of safety while pressing expansionist objectives.

If the new administration entered office intending to move cautiously in US–Soviet relations, events would soon bring Bush and Gorbachev together. During the last months of 1989, the

Soviet Eastern European empire began to unravel, forcing key Washington officials to reconsider their attitude towards Moscow. After weeks of temporizing, Bush, in early December 1989, journeyed to the Malta summit with Gorbachev realizing that any success in achieving Western goals depended on Gorbachev. During their sessions, Bush and Gorbachev defined an ambitious programme for cooperation and speedy progress on pending major arms-control measures and a Washington summit in late spring 1990.

Gorbachev arrived in Washington on 30 May 1990, beleaguered and profoundly unpopular at home but determined to sustain his image of confidence, enthusiasm, and authority. He and Bush signed protocols on verification of nuclear testing that had been agreed to nearly two decades earlier—the Threshold Test Ban Treaty (1974) and Peaceful Nuclear Explosions Treaty (1976)—that finally permitted their ratification. But Bush baulked at considering a comprehensive nuclear test ban; indeed, he had issued a policy statement earlier in January that his administration had 'not identified any further limitations on nuclear testing… that would be in the United States' national security interest'. The two leaders, nevertheless, pressed ahead with other issues, establishing a framework that would lead to future arms reductions—the Treaty on Conventional Forces in Europe (CFE) in November 1990 and the Strategic Arms Reductions Treaty (START I) in August 1991.

Accelerated discussions between 1989 and 1990—CFE negotiations had limped on for nearly twenty years—finally resolved most of the major areas of disagreement, and pointed towards a CFE agreement that a few years earlier would have been unthinkable. France joined the talks; Gorbachev unilaterally withdrew troops and equipment from forward areas; the Warsaw Pact disintegrated; and a reunited Germany agreed to troop limitations. Bush signed the CFE pact in Paris on 19 November 1990, placing limits on tanks, artillery, armoured combat vehicles, aircraft, helicopters, and military personnel stationed in Europe from the Atlantic Ocean to the Ural Mountains. In the history of arms control, this treaty was certainly a most impressive accomplishment. Before the signatures were dry on the CFE treaty, Soviet forces were already committed to a complete pull-out from Hungary and Czechoslovakia in 1991 and from the eastern parts of Germany in 1994; of the original twenty-three states involved in the CFE negotiations, one—the German Democratic Republic—had ceased to exist, and five others had changed their names, a clear indication of their independence from Moscow. With the Warsaw Pact disappearing, a US delegate to the CFE negotiations later claimed the treaty had 'ended the Cold War' since Europe was now truly at peace.

Bush–Gorbachev: START I

In January 1989, the new Bush White House found the basic framework of START already largely spelled out by the previous administration; however, because of the incessant bickering over technicalities between Reagan administration agencies several details had been left unresolved and the proposed treaty remained dormant. A major stumbling block centred on the issue of 'de-MIRVing'—how to reduce the number of warheads placed atop a single intercontinental missile. With newer generations of ICBMs, this could easily be achieved as monitoring of the testing and deployment of each other's new ICBMs would determine the number of counted MIRVs verified on each missile. A cheaper, faster method of de-MIRVing, initially proposed by Washington in 1987, would be to 'download' or remove warheads from existing ICBMs. This would allow the US to meet START limits by removing two of three warheads from the Minuteman III.

Another major hurdle was how to verify whether sea-launched cruise missiles (SLCMs) were physically aboard warships. American naval officers rebelled at the idea of Soviet inspectors snooping about their newest nuclear submarines. Tossing aside

Reagan's often repeated Russian proverb, 'trust but verify', the US finally proposed that each side 'declare' the number of SLCMs it planned to deploy. Dissatisfied with this solution, in a historic shift of roles Moscow now pressed for intrusive verification procedures. This demand unsettled Pentagon officials and US intelligence agencies that rebelled at the thought of Soviet inspectors prowling US defence factories and other installations.

At Moscow on 31 July 1991, after eight and a half frustrating years, Presidents Bush and Gorbachev signed a detailed 750-page START I treaty. The Americans finally conceded on the downloading issue, provided the total number of Soviet warheads was limited to 1,250, but at the last minute insisted that the Soviets test their three-stage SS-25 at 11,000 kilometres to ensure that it did not carry three warheads. As it emerged, the treaty would limit each side to the deployment of 1,600 ballistic missiles and long-range bombers, carrying 6,000 'accountable' warheads by 5 December 2001, and established further sublimits—4,900 warheads on deployed ballistic missiles, including no more than 1,100 warheads on deployed mobile ICBM systems, and 1,100 'accountable' bomber weapons. This was the first agreement that called for each side to reduce its strategic arsenal significantly, as some 25–35 per cent of the nuclear warheads carried on ballistic missiles were to be eliminated. In addition, START I incorporated the earlier INF Treaty's verification system that provided access to telemetry data and permitted on-site inspections. The US and Russia completed their START I reductions by 4 December 2001, and waited for Belarus, Kazakhstan, and Ukraine to turn over to Russia the former Soviet strategic nuclear weapons based in their territories.

Bush-Yeltsin: START II

In his State of the Union speech, on 28 January 1992, Bush announced that the US would unilaterally eliminate further work on the single-warhead Midgetman missile, and would assign a significant number of its bomber fleet to conventional roles. Additionally, he called for a START II agreement that could further reduce the number of American and Russian warheads, especially if his de-MIRVing initiative was accepted. Meanwhile, the new Russian president, Boris Yeltsin, proposed even deeper cuts that would reduce the number of warheads to 2,000 to 2,500 and, in a letter to Bush, denounced all MIRVs as 'the root of evil—from the point of view of threats to stability'. Washington perceived his proposal as radically affecting its traditional strategic triad.

A major problem standing in the way of proceeding rested with the return of the former Soviet Union's strategic nuclear weapons from Belarus, Kazakhstan, and Ukraine, but this was resolved with the Lisbon Protocol, signed on 23 May 1992. The protocol created a five-state START I regime, involving Belarus, Kazakhstan, Russia, Ukraine, and the United States, that called for all former Soviet strategic nuclear weapons to be returned to Russia. With this problem out of the way, serious negotiations began on START II. While Moscow pressed for substantial reductions, in Washington the Defense Department refused to accept the 2,500-warhead limit. As a concession, the Russians proposed a phased reduction with both sides beginning at 4,500 to 4,700 as the initial step, and later dropping to 2,500 by the year 2005.

At a two-day June 1992 meeting in Washington, Yeltsin provided the approach needed to complete the START II pact. Rather than trying to apply a numerical ceiling, he suggested using a range: in phase one, each side would have 4,250 to 3,800 warheads; in phase two, the range would decrease to 3,000 to 3,500. This satisfied the Russians for economic reasons, and the Pentagon for its force structure. Yeltsin recognized, according to Secretary Baker, 'that in the realm of nuclear weapons, a few-hundred-warhead advantage, when both sides had over three thousand warheads, was not all that important'. Bush concurred. The first phase would last until the year 2000; the second phase would limit warheads to 3,000 to 3,500 three years later. As final details of the treaty were being worked out, discussion arose as to whether Bush should continue with the process and the signing of the new START II, since Bill Clinton had already been elected to succeed him. A senior adviser later reported: 'It was feared that the Russians would simply take whatever concessions we offered, delay until after 20 January, and pick up with a new team.' President Bush made the decision to go for an agreement before he left office.

On 3 January 1993, Presidents Bush and Yeltsin signed START II. Since it relied heavily on START I for definitions, procedures, and verification, it could not enter into force until the START I ratified process was completed by 1994. Subsequently, the US ratified START II on 26 January 1996, and Russia on 14 April 2000. At the Helsinki Summit of March 1997, Presidents Bill Clinton and Boris Yeltsin agreed to extend the time for final reductions to 2007. Nevertheless, START II did not enter into force, because the US failed to ratify the 'Agreed Statements' that Presidents Bill Clinton and Boris Yeltsin signed in September 1997, a condition laid down by the new Russian President Vladimir Putin when endorsing Russian ratification in May 2000. Subsequently, Russia repudiated START II on 14 June 2002, the day following President George W. Bush's unilateral abrogation of the ABM Treaty.

Recall of tactical nuclear weapons

Tens of thousands of tactical nuclear weapons were deployed during the Cold War; however, leaders in the US and Soviet Union/Russia began the process of retrieving many of these weapons. Reciprocal unilateral pledges, initiated by President George H. W. Bush in September 1991, near the end of the Cold War, known collectively as Presidential Nuclear Initiatives (PNIs), succeeded in removing 'battlefield' nuclear weapons, such as nuclear artillery shells, from foreign deployment. Washington, vitally concerned about whether Moscow would be able to maintain control over its tactical nuclear weapons as the Warsaw Pact collapsed, hoped Kremlin leaders would follow suit. In his 27 September announcement, Bush specifically committed to withdraw 'all ground-launched short-range weapons deployed overseas and destroy them along with existing US stockpiles of the same weapons and cease deployment of tactical nuclear weapons on surface ships, attack submarines, and land-based naval aircraft during normal circumstances'. Implicitly, the United States reserved the right to redeploy these arms in a crisis.

A month later, Soviet President Mikhail Gorbachev responded with his own unilateral reciprocal measures. On 5 October, Gorbachev pledged to: eliminate all nuclear artillery munitions, nuclear warheads for tactical missiles, and nuclear mines; remove all tactical nuclear weapons from surface ships and multipurpose submarines (these weapons would be stored in central storage sites along with all nuclear arms assigned to land-based naval aircraft); and separate nuclear warheads from air defence missiles and put the warheads in central storage. A 'portion' would be destroyed. Subsequently, on 29 January 1992, Russia's president-elect Boris Yeltsin agreed to uphold Gorbachev's commitments, and declared Russia would further eliminate a third of its sea-based tactical nuclear weapons and half of its ground-to-air nuclear missile warheads, while halving its airborne tactical nuclear weapons stockpile. Pending reciprocal US action, the other half of this stockpile would be taken out of service and placed in central storage depots.

The Lisbon Protocol of 31 July 1991 promptly extended recognition to four republics of the former Soviet Union as designated successor states committed to various earlier arms-control treaties such as START I. The reason for the urgency was that these republics held portions of the Soviet Union's strategic nuclear weapons and the area was threatened with the possibility of civil wars; yet, no mention was then made of who controlled the smaller tactical nuclear weapons, although they were more numerous and more widely spread than the strategic weaponry.

The republics quickly understood that they could not keep these weapons in their own arsenals, for attempts to take over control of nuclear weapons would provide Russia with a pretext to invade. Washington left no doubt that it desired control of the nuclear weapons to remain in Moscow's hands and used its influence to encourage the new states to allow Russia to withdraw the tactical nuclear weapons. Accordingly, Moscow launched 'a speedy and largely secret' programme to extract all of them by May 1992; some 3,000 tactical nuclear weapons were removed from Ukraine alone. The prospects for elimination of tactical nuclear weapons have remained dormant since the early 1990s primarily because of considerable discrepancy between the smaller American inventory and the much larger imputed Russian stockpile.

During the Cold War, the US had sent some 5,000 tactical nuclear weapons overseas, most of which were assigned to NATO. By the end of 1992, the US had completed its pledged reductions and withdrawals; a year later it had destroyed nearly 3,000 tactical nuclear weapons. The Soviet/Russian stockpile was thought to range from 12,000 to nearly 21,700 weapons. Determining US and Soviet/Russian fulfilment of their PNI obligations, however, was difficult—then and now—because of the ambiguity related to the composition, size, and location of these weapons. After confirming that all Soviet tactical nuclear weapons had been returned from the four republics of the former Soviet Union, Moscow announced in May 2005 that these arms 'are now deployed only within the national territory and are concentrated at central storage facilities of the Ministry of Defense'.

A precise assessment of the US and Russia's tactical nuclear stockpiles, therefore, is difficult since recent statements vary so much. Indeed, one account reports that US-NATO forces retain hundreds of tactical nuclear weapons in Europe, and the Russian arsenal contains a much higher number. Estimates are that the US retained nearly 1,100 tactical nuclear warheads, some 480 of which are nuclear gravity bombs stockpiled in Belgium, Germany, Italy, the Netherlands, Turkey, and the United Kingdom. Russia is estimated to be storing 3,000 to 6,000 nonstrategic weapons. While Russia opposed the stockpiles of nuclear gravity bombs in Europe and the US lamented the lack of Russian transparency regarding its tactical arsenal, the two nations have not seriously sought negotiations to gain further reductions of tactical nuclear weapons since the early 1990s.

A further factor confusing the issue is that many tactical nuclear weapons are, in fact, dual-use weapon systems that can be fitted with different types of warheads, including nuclear, high explosive conventional, biological, or chemical warheads. The wide proliferation of these delivery vehicles—without special warheads—during the Cold War has left a complicated legacy of wide dispersal of dual-use delivery vehicles around the world. Short-range, dual-use rockets and missiles were dispersed by the Soviets to precisely the areas of greatest proliferation threat in more recent years: Iran, North Korea, Iraq, Egypt, Yemen, Afghanistan, Pakistan, Kazakhstan, Turkmenistan, Vietnam, and Belarus, among other countries. Many of those same countries, including Cuba, have also had FROG (Free Rocket Over Ground) rockets. Various American-developed dual-use systems were widely dispersed through Western Europe and Israel, in particular. Several of these countries have taken the original designs and improved on them to build new generations of missiles. Consequently, possession and possible use of tactical or short-range nuclear weapons in regional conflicts remains an issue in the 21st century.

A continuing NATO controversy

In 2010, as NATO defence and foreign ministers met to review the alliance's draft 'Strategic Concept', several analysts called for a comprehensive review of what they saw as an outdated nuclear policy. In the October issue of *Arms Control Today*, Oliver Meier and Paul Ingram reported that the twenty-eight members of NATO were divided regarding the future role of nuclear weapons in the alliance's defence policy and urged reconsideration of the Strategic Concept. Later, at a conference addressing 'Next Steps in Arms Control', Meier noted that NATO's review of the Strategic Concept should take place in the context of President Obama's global zero nuclear policy since it had the support of a broad majority of European parliaments and publics. Regretfully, he observed, NATO's current policy 'still is based on the Cold War theory that short-range nuclear weapons could be used to defeat conventional superior Soviet forces'.

If several European governments, including at least three of the five nations stockpiling the US tactical nuclear weapons, desired their withdrawal, several Central European nations and Turkey had reservations related not nearly as much to the military value of these weapons as to the credibility of the US's and NATO's security assurances. Their concern centred on Russia's stationing of tactical nuclear weapons near its borders with NATO states, and its emphasis on nuclear readiness, not to mention Moscow's renunciation of its long-held 'no-first-use' of nuclear weapons—an admission of Russia's weakened conventional arsenal.

Cooperative Threat Reduction (CTR) programmes

The sudden demise of the Soviet Union and the chaos that followed led, in 1992, to the US's CTR programme, usually called the Nunn–Lugar programme after its sponsors Senators Sam Nunn (D-GA) and Richard Lugar (R-IN). The Nunn–Lugar legislative Act of 12 November 1992 provided US financial assistance to the Commonwealth of Independent States, especially Russia, to consolidate the former Soviet nuclear arsenal and ensure their custodial safety. Later, assistance in the destruction of former Soviet chemical arms was added to the programme. The ten-year, four-billion-dollar programme was designed to:

- 'destroy nuclear weapons, chemical weapons, and other weapons',
- 'transport, store, disable, and safeguard weapons in connection with their destruction', and
- 'establish verifiable safeguards against the proliferation of such weapons'.

It was a bargain. Or as a supporter of the original budget pointed out, the cost of the programme, some $400 million per year, has been less than half of what Americans spend annually on cat food.

In two decades of CTR efforts, much has been accomplished: 7,519 nuclear warheads deactivated; 768 ICBMs, 498 ICBM silos, 148 mobile ICBM launchers, 651 SLBMs, 476 SLBM launchers, 32 ballistic missile-capable submarines, 155 strategic bombers, 906 air-to-surface missiles, and 194 nuclear test tunnels destroyed; 24 security upgrades implemented at nuclear weapon storage sites; and 469 train shipments of nuclear weapons moved to more secure, centralized storage sites. The programme purchased 500 metric tons of highly enriched uranium from dismantled warheads and helped remove all nuclear weapons from Ukraine, Kazakhstan, and Belarus—states that once held the world's third, fourth, and eighth largest nuclear arsenals. Additionally, nineteen biological agent monitoring stations have been established.

Global Threat Reduction Initiative (GTRI)

Established in May 2004, the GTRI is a collaborative programme aimed at securing vast stocks of dangerous nuclear material scattered around the world. Where the CTR's focus is on nuclear weapons material in states of the former Soviet Union, the GTRI is a complementary programme involved in 'repatriating or otherwise securing nuclear fuel' from peaceful use facilities and converting these facilities 'to use new, more proliferation-resistant technology'. The GTRI has been made necessary because of efforts by programmes such as Atoms for Peace, the International Atomic Energy Agency, and the Nuclear Non-proliferation Treaty to assist non-nuclear weapons states to gain peaceful nuclear technology.

The GTRI programme, which pulled together several existing programmes in the Department of Energy, attempted to secure highly enriched uranium (HEU) that could be used to construct a primitive nuclear weapon. In addition to improving HEU security, the programme sought to convert reactors to use low-enriched uranium (LEU) fuel that cannot be used for bomb-making. A successful programme that secured medical nuclear isotopes made from HEU reduced the prospect of terrorists using them for radiological or 'dirty' bombs. Thus, GTRI reported, it had 'secured more than 960 radiological sites around the world containing over 20 million curies, enough for thousands of dirty bombs', 'removed more than 120 nuclear bombs' worth of highly enriched uranium and plutonium, [and] secured more than 775 bombs' worth of HEU and plutonium' from a reactor in Kazakhstan. Since 2004, twenty-two research reactors have been converted to using LEU fuel and twelve other HEU research reactors have been shut down. The efforts to return HEU fuel to its origins has resulted in thirty-five shipments to Russia of more than 1,490 kilograms of Russian-origin HEU and to America more than 320 kilograms of US-origin HEU from many international partners—including Australia, Germany, Austria, Greece, Japan, Argentina, Sweden, Portugal, Romania, Taiwan, and the Netherlands.

The Strategic Offensive Reductions Treaty (SORT)

American President George W. Bush's administration frequently demonstrated a dislike for traditional arms-control agreements.

This was clearly evident in the withdrawal from the ABM Treaty and continued lack of interest in the comprehensive test ban treaty, retention of the nuclear first-use policy, and a reliance on nuclear weapons.

Following the terrorist events of 9/11, the role of nuclear strategists in Defense Secretary Donald Rumsfeld's Pentagon came to dominate. For example, the administration's early 2002 secret Nuclear Posture Review instructed the Pentagon to draft contingency plans for the use of nuclear weapons against at least seven countries, naming not only Russia and the 'axis of evil'—Iraq, Iran, and North Korea—but also China, Libya, and Syria. It reflected the fundamental inconsistency between America's diplomatic objectives of reducing nuclear arsenals and preventing the proliferation of weapons of mass destruction, on the one hand, and the military imperative to prepare for the unthinkable, on the other.

George W. Bush entered the White House vowing to reduce America's nuclear weapons, as he told the National Press Club on 23 May 2000, to 'the lowest possible number consistent with our national security'. Bush had initially sought to implement strategic offensive weapons reductions with unilateral declarations and a handshake, but Russian President Vladimir Putin insisted upon a more formal arrangement. Putin desired stipulated reductions that would provide a sense of parity, predictability, and reduced expenditures that could be accomplished by reducing each side to 1,500 warheads. The new administration's aversion to formal arms-control agreements, coupled with its enthusiasm for unilateral actions, resulted in SORT, signed by Presidents Bush and Putin at Moscow on 24 May 2002 and entering into force on 1 June 2003. The brief treaty, however, focused more on limits in the tradition of the early Cold War treaties rather than reductions that had been the hallmark of START I and II. SORT ignored the agreement in principle reached by Presidents Clinton and Yeltsin as an outline for START III that stipulated cuts to their strategic arsenals to 2,000 to 2,500 warheads each, while requiring significant reductions in delivery vehicles.

While in SORT, also known as the Moscow Treaty, the United States and Russia each agreed to deploy no more than 1,700 to 2,200 strategic warheads by the end of 2012—when the treaty expired—the pact did not restrict the number of permitted delivery vehicles that could be retained so long as neither party exceeds the START I limits. Nor did the treaty require the destruction of delivery vehicles or agree on specific counting rules; hence, a MIRVed nose cone might contain a single warhead and be counted as a single warhead, although it could quickly be loaded with nine additional warheads. Warheads in excess of the stipulated limits did not have to be dismantled or destroyed; they could simply be stored. Thus, the Bush administration stated that it planned to maintain at least 2,400 warheads in a ready-reserve status. According to a pessimistic assessment by arms-control specialists, the treaty totals less than 500 words, repudiates key arms-control principles and achievements, eschews predictability, and compounds the proliferation dangers from Russia's unsecured nuclear weapons complex. SORT, in essence, regarded each nation's nuclear programme to be its own business.

The new START

The Bush administration fumbled attempts to devise a successor agreement before START I expired. A central element of START I, signed by the United States and the Soviet Union in July 1991, was that both parties had used its verification system to monitor the SORT pact. Knowing how long treaty negotiations may take, Bush officials could have made reaching some agreement to extend the verification system a priority, but they did not. As the deadline had passed, the Barack Obama administration was faced with the daunting challenge of negotiating a completely new treaty that could win the votes of sixty-seven US senators, while also dealing with Russian objections to proposed US missile defences in

Europe that had not been a consideration when START I was first contemplated. The Obama administration was well aware of Russia's concerns—repeated again and again during the Bush years by President Putin and by Russian defence ministers and generals. But in 2009 and 2010 the stakes were even higher. Bush had only desired Russian acquiescence to his plan to deploy US missile defence systems in Poland and the Czech Republic; Obama desired that too, but his administration also needed Russian cooperation on a new START I treaty.

This history was reflected anew in the negotiations for a replacement to START I. Despite early setbacks in the ongoing talks, Presidents Barack Obama and Dmitry Medvedev, following an 18 December 2009 meeting at Copenhagen, were optimistic that a new pact would be ready in the near term. Stalling the agreement was Washington's demand for unencrypted telemetry data from US and Russian offensive missile tests to be shared, as was required under the original 1991 START I agreement. Moscow, in turn, was now coupling US desire for unencrypted telemetry data to more data about US missile defences. The Americans sought, as agreed in START I, telemetric data after each flight test, together with the key to interpreting the data, and a pledge not to jam or encrypt such data. The stumbling block occurred because the US was not building new ICBMs but rather was upgrading current models, such as the Trident D-5. Meanwhile, the Russians planned to test and deploy new missiles, such as the RS-24 mobile missile, as replacements for aged Soviet-era ones. Thus, the US would have no offensive missile tests to report, but the Russians would; however, the US would be testing anti-missile interceptors but did not want to be required to share these data with Moscow. On 29 December 2009, Prime Minister Vladimir Putin declared on television that Russia needed more detailed information about US missile defences. Concerned that missile defence would give the US an advantage, Putin explained, 'The problem is that our American partners are developing missile defenses, and we are not.'

Disagreement over strategic delivery systems also posed a challenge. Moscow repeatedly sought lower numbers than Washington preferred. In July 2009, Obama and Medvedev had suggested that limits would range between 500 and 1,100; later the Americans focused on a middle ground of around 800. This was near the number of currently deployed US delivery systems. Russia, which currently deployed only some 620 systems, pressed for a lower figure of about 550. At whatever level delivery systems were set at, it was expected that the number of warheads would be limited to around 1,600. Meanwhile, all forty Republican senators, plus Independent Joseph Lieberman, warned President Obama on December 15 that: '[W]e don't believe further reductions can be in the national security interest of the US in the absence of a significant program to modernize our nuclear deterrent.' The administration was placed on notice that ratification of any new treaty was going to come with a price tag since a new pact required approval by two-thirds of the Senate.

Obama and Medvedev: New START

On 8 April 2010 in Prague, Obama and Medvedev signed the 'New START' treaty that would replace the expired 1991 START I agreement. The legally binding, verifiable pact limited each nation's deployed strategic nuclear warheads to 1,550 and strategic delivery systems to 800 deployed and nondeployed—both significant reductions. This meant the treaty-accountable warhead limit would be 30 per cent lower than the 2,200 limit of SORT, and the delivery vehicle limit 50 per cent lower than the 1,600 allowed in START I. The new 1,550 limit applied to deployed nuclear warheads on ICBMs, SLBMs, and heavy bombers carrying a single nuclear warhead. The new agreement provided each side with the freedom to mix and to develop its own force structure. Although each nation was limited to 800 strategic delivery systems, only 700 were allowed to be deployed with the others used for training and testing. Launchers without missiles were counted as nondeployed. Reductions required by the pact were to be carried out within seven years after the treaty entered into force. Additionally, the agreement tailored elements of the verification regime from START I, often streamlining monitoring provisions to meet the requirements of the New START era. These measures included on-site inspections and exhibitions, data exchanges and notifications related to strategic weaponry contained in the treaty, together with enhanced verification monitoring via national technical means. The provisions for inspections provided for two types: 'Type One' that takes place at ICBM, submarine, and air bases, while 'Type Two' targets other facilities such as ICBM loading areas, test ranges, and training sites. A major reason for easing the on-site inspections was the significant advances over the past two decades in the ability to verify compliance with national technical means. These enhancements made it possible to gain much more relevant information than was the case in 1991.

Ratification of New START came much easier in Moscow than in Washington. The lower house of Russia's parliament gave overwhelming preliminary approval of the treaty in a 350–58 vote in late December 2010 and the next month approved the second and third reading. A day after, the upper house concurred in late January 2011 and, when signed by Medvedev, the ratification procedure was completed. In Washington, meanwhile, various Republican senators raised several questions: Would the president allocate sufficient funds for modernization of the US nuclear forces? Did the treaty interfere with the US's planned deployment of missile defences? And why did the treaty not reduce tactical nuclear weapons? Following an eight-month delay and eight days of debate, the Senate on December 22 voted 71–26 to ratify the treaty that entered into force on 5 February 2011.

Approval was accomplished at a price, however, as the administration pledged $10 billion over ten years to increase an already enlarged budget for the nuclear weapons complex. Given budget constraints, the US will undoubtedly reconsider the promise of additional modernization funding and perhaps even its more than forty-year-old 'triad' system of ICBMs, SLBMs, and bombers.

Zero nuclear weapons?

In Prague on 5 April 2009, President Obama outlined a path towards 'a world without nuclear weapons'. The hope of enhancing the world's safety by abolishing nuclear weapons certainly was not a new idea. If Obama was not the first to promote the idea of zero nuclear weapons, as president of the remaining superpower he bestowed upon the notion of eliminating nuclear weaponry a particularly significant blessing that drew worldwide attention and approval. And in doing so, he directed attention to elements of the non-proliferation regime 'to cut off the building blocks needed for a bomb'. Working together, he said, 'we will strengthen the nuclear Non-Proliferation Treaty as a basis for cooperation'. To this end, he pledged to pursue ratification of the Comprehensive Test Ban Treaty, seek a new treaty that would verifiably halt the production of fissile materials, and find a new framework for civil nuclear cooperation, including a new international fuel bank, so that countries can access peaceful power without increasing the risks of proliferation.

Accordingly, the International Atomic Energy Agency's board, in December 2010, approved a fuel bank plan, eligibility for which depends on a state agreeing to comprehensive safeguards to all its peaceful nuclear activities. The United States, the European Union, Kuwait, the United Arab Emirates, and Norway pledged $100 million to purchase and take delivery of some 60–80 metric tons of LEU for the bank. Finally, the president's message recognized that not all nations would follow the rules. 'We go forward with no illusion. Some will break the rules, but that is why we need a structure in place that ensures that when any nation does, they will face consequences.' He went on to stress the need to 'strengthen international inspections' and 'real and immediate consequences'. Building the bomb was always going to be easier than finding the means of harnessing it.

Trump and Iran

On 8 May 2018, President Donald Trump made good on his campaign promise, and announced he was withdrawing from the Joint Comprehensive Plan of Action, the agreement that established a set of robust verifiable limits on Iran's ability to develop a nuclear weapon for at least the next ten or fifteen years. Dismantling the signature foreign policy achievement of President Barack Obama—reached in Vienna on 14 July 2015, between Iran, P5+1 (the five permanent members of the United Nations Security Council—China, France, Russia, the United Kingdom, and the United States—plus Germany), and the European Union—Trump reimposed the stringent sanctions the US had applied to Iran before the deal, enforcing his 'maximum pressure' campaign to rein in Tehran's nuclear and regional ambitions. The original nuclear deal had tightly restricted Iran's nuclear ambitions in exchange for ending sanctions that had crippled its economy.

The other signatories to the deal said that they would remain in the agreement. But Trump put allies on notice that European countries would face American sanctions if they did business with Iran and would have to choose between the United States and the Islamic Republic of Iran. Such is the force of US secondary sanctions that few European nations would be able to resist. Iran, desperate and running out of options, has responded with attacks on tankers in the Persian Gulf and the Gulf of Oman, astride the strategic Strait of Hormuz, through which 20 per cent of the world's oil flows, and shooting down an American surveillance drone over disputed territory, while ramping up its production of nuclear fuel, following through on its threat to walk away from the nuclear deal. The stage is set for the United States and Iran, two long-time adversaries, hurtling towards potential crisis and war.

Meanwhile, Iran's historical enemies, Israel and Saudi Arabia, wait and watch.

Trump–Kim negotiations

After a war of fiery words following North Korean violations of United Nations sanctions (Figure 8), United States President Trump and North Korea's Kim Jong-Un met in Singapore on 12 June 2018, for the first summit between the leaders of the US and North Korea since the end of the 1950–3 Korean War, to resolve the nuclear crisis. After a spectacular display of personal summit diplomacy, on both sides, Trump and Kim agreed to the 'complete denuclearization' of the Korean Peninsula, but without determining what that meant or how to achieve it. Discussion on the finer details was saved for another day. In the meantime, Kim continued his suspension of nuclear and long-range missile tests, while Trump persisted with sanctions and pressure against North Korea.

8. North Korean leader Chairman Kim observing a ballistic missile test.

In February 2019, Trump and Kim met for a second time, in Hanoi, Vietnam, the meeting breaking down after a rift emerged over the lifting of sanctions. Specifically, Trump rejected Kim's call for major sanctions relief in return for dismantling his main nuclear complex. Several months later, in June 2019, during their third encounter, the president shook hands with Chairman Kim at a jointly controlled area inside the Demilitarized Zone between the two Koreas, becoming the first US president to cross over into North Korea. During a brief discussion the two leaders agreed that working-level talks should resume. After three summits, neither side had much to show for its efforts, save the fact that North Korea continued to conduct short-range missile tests, a programme that has drawn little criticism from Washington, which has tended to focus upon ensuring a moratorium on Pyongyang's nuclear weapons and long-range ballistic missile testing. Renewed tensions on the Korean peninsula followed the collapse of the denuclearization negotiations.

Trump and Putin

In February 2019, the Trump administration announced that the US would begin withdrawing from a landmark 1987 nuclear arms pact with Russia—the INF Treaty—asserting that the Kremlin had violated it for years. Because of their short flight times—as little as ten minutes—the intermediate-range missiles had come to be seen as a hair trigger for nuclear war, at worst, and a constant menace to America's NATO allies in Europe at the very least. According to information dating back to the Obama administration, the US laid responsibility for the breach at the feet of Russian President Vladimir Putin, who denied all allegations, while boasting of the growing threat of a new class of hypersonic weapons.

While Trump and Secretary of State Mike Pompeo have called the INF Treaty outdated—with the green light already given to a new class of American intermediate-range missiles—European leaders say the answer is to renegotiate the treaty, not to scrap it. Russian hawks, seeing one more opportunity to drive a wedge between Europe and Washington, have emphasized the risks of a US nuclear missile build-up. Despite the apparent personal chemistry between Trump and Putin, the Russian–American nuclear rivalry has an ominous feeling about it.

The nuclear rivalry with Russia, North Korea, and Iran gives the sense that we are still living in the Cold War period. The Cold War simply took a break for thirty years, from the fall of the Berlin Wall in 1989 to the present. Whatever we call it—the new Cold War or Cold War II—there is again a sharp confrontation between nuclear-armed powers, with the ever present threat of a deliberate or unintended military confrontation between Russia and NATO members, which has not, at least up to now, turned into Armageddon.

References and further reading

The publications listed below represent what I consider to be the most appropriate sources of further reading for newcomers to a subject for which there is a vast literature. Constraints of space have made it necessary to omit many excellent and otherwise important works in the field.

Preface

Congressman Adam Smith's views on the perils of a new nuclear arms race are found in *Arms Control Today*, 48:10 (December 2018), 6–9. Les Aspin, Winston Churchill, and Madeleine Albright's comments are located, respectively, in David G. Coleman and Joseph M. Siracusa, *Real-World Nuclear Deterrence: The Making of International Strategy* (Praeger Security International, 2006); and Joseph M. Siracusa, 'The "New" Cold War History and the Origins of the Cold War', *Australian Journal of Politics and History*, 47:1 (2001), 149–55.

Chapter 1: What are nuclear weapons?

The best place to begin the study of nuclear weapons is the pages of the *Bulletin of Atomic Scientists*, founded in 1945 as a newsletter distributed among nuclear physicists concerned by the possibility of nuclear war; for seventy-five years the *Bulletin*'s iconic Doomsday Clock has followed the rise and fall of nuclear tensions. For a discussion of the far-reaching effects of early Cold War nuclear weapons testing, see David M. Blades and Joseph M. Siracusa, *A History of U.S. Nuclear Testing and*

Its Influence on Nuclear Thought, 1945–1963 (Rowman & Littlefield, 2014); and John R. Walker, *British Nuclear Weapons and the Test Ban, 1954–1973. Britain, the United States, Weapons Policies and Nuclear Testing: Tensions and Contradictions* (Ashgate, 2010). For an introduction to the threat posed by nuclear terrorists, see Graham Allison, *Nuclear Terrorism: The Ultimate Preventable Catastrophe* (Times Books, 2004). Also useful are Scott D. Sagan and Kenneth N. Waltz, *The Spread of Nuclear Weapons: A Debate Renewed* (W. W. Norton, 2003); Joseph Cirincione, Jon Wolfstahl, and Miriam Rajkumar, *Deadly Arsenals: Nuclear, Biological and Chemical Threats* (Carnegie Endowment for International Peace, 2005); and Alethia H. Cook, *Terrorist Organizations and Weapons of Mass Destruction: U.S. Threats, Responses, and Policies* (Rowman & Littlefield, 2017).

Chapter 2: Building the bomb

For details on the various nuclear activities of World War II contestants, see Richard Dean Burns and Joseph M. Siracusa, *A Global History of the Nuclear Arms Race: Weapons, Strategy, and Politics* (2 vols, Praeger, 2013). Also useful are Richard Rhodes, *The Making of the Atomic Bomb* (Simon and Schuster, 1986); McGeorge Bundy, *Danger and Survival: Choices about the Bomb in the First Fifty Years* (Random House, 1988); and Mark Walker, *German National Socialism and the Quest for Nuclear Power, 1939–1949* (Cambridge University Press, 1989). The story of Einstein's famous letter to President Franklin D. Roosevelt is ably told in Walter Isaacson, *Einstein: His Life and Universe* (Simon & Schuster, 2008); and Martin J. Sherwin, *A World Destroyed: Hiroshima and Its Legacies* (3rd edn, Stanford University Press, 2003).

The 'atomic diplomacy' debate may be followed in Gar Alperovitz, *The Decision to Use the Atomic Bomb and the Architecture of an American Debate* (Harper Collins, 1995); Robert James Maddox, *Weapons for Victory: The Hiroshima Decision Fifty Years Later* (University of Missouri Press, 1995); and Wilson D. Miscamble, *The Most Controversial Decision: Truman, the Atomic Bomb, and the Defeat of Japan* (Cambridge University Press, 2011).

For the impact of wartime bombing of civilians, see Jorg Friedrich, *The Fire: The Bombing of Germany, 1940–1945* (Columbia University Press, 2007); and the incomparable John Hersey, *Hiroshima* (Penguin, 1946).

Chapter 3: A choice between the quick and the dead

For a detailed treatment of the Baruch Plan, which contains a fair amount of primary source material, see Leneice N. Wu's essay in Richard Dean Burns (ed.), *Encyclopaedia of Arms Control and Disarmament* (3 vols, Charles Scribner's Sons, 1993); and Richard G. Hewlett and Oscar E. Anderson, Jr, *A History of the United States Atomic Energy Commission*, vol. 1, *The New World, 1939/1946* (University of Pennsylvania, 1962).

The best historical assessments include Barton J. Bernstein, 'The Quest for Security: American Foreign Policy and International Control of Atomic Energy', *Journal of American History*, 60 (March 1974), 1003–44; and Larry Gerber, 'The Baruch Plan and the Origins of the Cold War', *Diplomatic History*, 6 (Winter 1982), 69–95.

Chapter 4: Race for the H-bomb

The H-bomb decision, together with its Cold War implications, may be followed in Coleman and Siracusa, *Real-World Nuclear Deterrence*; and Ken Young and Warner Schilling, *Organizational Conflict and the Development of the Hydrogen Bomb* (Cornell University Press, 2020). The famous NSC 68 document is reassessed in Ken Young, 'Revisiting NSC 68', *Journal of Cold War Studies*, 15:1 (Winter 2013), 3–33. Comments by the Atomic Energy Commission's advisory committee against, and Truman's defence of, the H-bomb are found in Blades and Siracusa, *A History of U.S. Nuclear Testing and Its Influence on Nuclear Thought, 1945–1963*.

For the historical context of these years, see Norman A. Graebner, Richard Dean Burns, and Joseph M. Siracusa, *America and the Cold War, 1941–1991* (2 vols, Praeger, 2010); and David James Gill, *Britain and the Bomb: Nuclear Diplomacy, 1964–1970* (Stanford University Press, 2014).

The Soviet side of this story is told ably in David Holloway, *Stalin and the Bomb: The Soviet Union and Atomic Energy, 1939–1956* (Yale University Press, 1994); and Vojtech Mastny, *The Cold War and Soviet Insecurity: The Stalin Years* (Oxford University Press, 1996). Also see, Vladislav M. Zubok and Constantine Pleshakov, *Inside the Kremlin's Cold War: From Stalin to Khrushchev* (Harvard University Press, 1996).

The story of the global anti-nuclear campaign, together with the forces, personalities, and events that moulded it, is told in Lawrence S. Wittner, *The Struggle Against the Bomb* (3 vols, Stanford University Press, 1993–2003).

Chapter 5: Nuclear deterrence and arms control

Indispensable are three works by Raymond L. Garthoff: *Deterrence and the Revolution in Soviet Military Doctrine* (Brookings Institution, 1990), *Soviet Strategy in the Nuclear Age* (revised edition, Praeger, 1962), and *Détente and Confrontation: American Soviet Relations from Nixon to Reagan* (Brookings Institution, 1985). In this same category I also include Lawrence Freedman's *The Evolution of Nuclear Strategy* (3rd edn, Palgrave Macmillan, 2003) and *Deterrence* (Polity, 2004); and Richard Dean Burns, *The Evolution of Arms Control: From Antiquity to the Nuclear Age* (Praeger Security International, 2009).

The treaty milestones of these years are covered in Richard Dean Burns (ed.), *Encyclopaedia of Arms Control and Disarmament* (3 vols, Charles Scribner's Sons, 1993).

Arms control efforts to limit the potential threat of strategic weaponry may be found in McGeorge Bundy, *Danger and Survival*; and J. P. G. Freeman, *Britain's Nuclear Arms Control Policy in the Context of Anglo-American Relations, 1957–68* (St Martin's Press, 1986).

The debates over nuclear testing and negotiation of the limited test ban treaty are detailed in Robert Divine, *Blowing on the Wind: The Nuclear Test Ban Debate, 1954–1960* (Oxford University Press, 1978); Glenn Seaborg, *Kennedy, Khrushchev and the Test Ban* (University of California Press, 1981); and Blades and Siracusa, *A History of U.S. Nuclear Testing and Its Influence on Nuclear Thought, 1945–1963.*

For the Reagan years and arms control, see James Mann, *The Rebellion of Ronald Reagan: A History of the End of the Cold War* (Viking, 2009); and Norman A. Graebner, Richard Dean Burns, and Joseph M. Siracusa, *Reagan, Bush, Gorbachev: Revisiting the End of the Cold War* (Praeger Security International, 2008).

For Soviet commentary on the nuclear arms control, see Anatoly Dobrynin, *In Confidence: Moscow's Ambassador to America's Six Cold War Presidents* (Times Books, 1995); and Mikhail Gorbachev,

Memoirs (Doubleday, 1995) and *Reykjavik: Results and Lessons* (Sphinx Press, 1987).

Box 1 Stages of weapons development, in Philip E. Coyle, *Arms Control Today*, 32 (May 2002): 5.

Box 2 Ballistic missile basics, in *Arms Control Today*, 31–4 (July/August 2002).

Chapter 6: Star Wars and beyond

Overviews include Richard Dean Burns and Lester H. Brune, *The Quest for Missile Defenses, 1944–2003* (Regina Books, 2004); and James M. Lindsay and Michael E. O'Hanlon, *Defending America: The Case for Limited National Missile Defense* (Brookings Institution Press, 2001). Also, see Steven J. Zaloga, *The Kremlin's Nuclear Sword: The Rise and Fall of Russia's Strategic Nuclear Forces, 1945–2000* (Smithsonian Institution Press, 2002).

The 1968 debates over deployment are covered in Edward R. Jayne, *The ABM Debate: Strategic Defense and National Security* (Center for Strategic Studies, 1969); Abram Chayes and Jerome Wiesner (eds), *ABM: An Evaluation of the Decision to Deploy an Antiballistic Missile System* (Harper and Row, 1969); and James Cameron, *The Double Game: The Demise of America's First Missile Defence System and the Rise of Strategic Arms Limitation* (Oxford University Press, 2018).

For Reagan's initiative, consult William L. Broad, *Teller's War: The Top Secret Story Behind the Star Wars Deception* (Simon and Schuster, 1992); Sydney Drell et al., *The Reagan Strategic Defense Initiative: A Technical, Political and Arms Control Assessment* (Harvard University Press, 1985); and Frances Fitzgerald, *Way Out There in the Blue: Reagan, Star Wars and the End of the Cold War* (Simon and Schuster, 2000). For the Soviet perspective, see David S. Yost, *Soviet Ballistic Missile Defense and the Western Alliance* (Harvard University Press, 1988).

Chapter 7: Post-Cold War era

For an overview of the post-Cold War history of nuclear non-proliferation, see Burns and Siracusa, *A Global History of the Nuclear Arms Race: Weapons: Strategy, and Politics*, vol. 2; and Joseph M. Siracusa and Aiden Warren, *Weapons of Mass*

Destruction: The Search for Global Security (Rowman & Littlefield, 2017).

The Soviet/Russian side of the story is ably told in Steven J. Zalogo, *The Kremlin's Nuclear Shield: The Rise and Fall of Russia's Strategic Nuclear Forces, 1945–2000* (Smithsonian Institution Press, 2002); Raymond L. Garthoff, *The Great Transition: American–Soviet Relations and the End of the Cold War* (The Brookings Institution, 1994); and Vladislav M. Zubok, *A Failed Empire: The Soviet Union in the Cold War from Stalin to Gorbachev* (University of North Carolina Press, 2007).

Notable insiders' accounts include: James Baker, III, *The Politics of Diplomacy: Revolution, War and Peace, 1989–1992* (Putnam's, 1995); Robert M. Gates, *From the Shadows: The Ultimate Insider's Story of Five Presidents and How They Won the Cold War* (Simon & Schuster, 1996); and James B. Goodby, *At the Borderline of Armageddon: How American Presidents Managed the Atom Bomb* (Rowman & Littlefield, 2006).

The zero nuclear weapons debate can be followed in Bruce G. Blair, *Global Zero Alert for Nuclear Forces* (The Brookings Institution, 1995); and George Perkovich, *Abolishing Nuclear Weapons: A Debate* (Carnegie Endowment for International Peace, 2009).

For an overall perspective, see William Walker, *A Perpetual Menace: Nuclear Weapons and International Order* (Routledge, 2012); Nicholas L. Miller, *Stopping the Bomb: The Sources and Effectiveness of US Nonproliferation Policy* (Cornell University Press, 2018); and Matthew Ambrose, *The Control Agenda: A History of the Strategic Arms Limitation Talks* (Cornell University Press, 2018).

Index

For the benefit of digital users, indexed terms that span two pages (e.g., 52–53) may, on occasion, appear on only one of those pages.

D

E

F

G

H

I

J

K

L

M

N

O

P

R

S

T

U

V

W

Y

Z

NUCLEAR POWER

A Very Short Introduction

Maxwell Irvine

The term 'nuclear power' causes anxiety in many people and there is confusion concerning the nature and extent of the associated risks. Here, Maxwell Irvine presents a concise introduction to the development of nuclear physics leading up to the emergence of the nuclear power industry. He discusses the nature of nuclear energy and deals with various aspects of public concern, considering the risks of nuclear safety, the cost of its development, and waste disposal. Dispelling some of the widespread confusion about nuclear energy, Irvine considers the relevance of nuclear power, the potential of nuclear fusion, and encourages informed debate about its potential.

www.oup.com/vsi